Xenobiotic Metabolic Enzymes: Bioactivation and Antioxidant Defense

Xenobiotic Metabolic Enzymes: Bioactivation and Antioxidant Defense

Chang-Hwei Chen

Xenobiotic Metabolic Enzymes: Bioactivation and Antioxidant Defense

Springer

Chang-Hwei Chen
Institute for Health and the Environment
and Department of Biomedical Sciences
University at Albany, State University
of New York, 5 University Place
Rensselaer, NY, USA

ISBN 978-3-030-41678-2 ISBN 978-3-030-41679-9 (eBook)
https://doi.org/10.1007/978-3-030-41679-9

This Springer imprint is published by the registered company Springer Nature Switzerland AG
The registered company address is: Gewerbestrasse 11, 6330 Cham, Switzerland

This book is dedicated to the loving memory of my wife
Julia Hsiu-Li Chen who still inspires me in everything I do

Preface

Xenobiotics are commonly referred to as foreign compounds that humans digest or inhale and are not produced by the body, such as foods, drugs, other chemicals, and air pollutants. Metabolic conversion of foreign compounds to facilitate their removal from the body is catalyzed by activation and detoxification enzymes. These enzymes are essential for bioactivation and detoxification of foreign compounds. Many xenobiotics are non-toxic in nature; however, some of them become cytotoxic or carcinogenic after metabolic conversion. Reactive intermediates or metabolites generated during metabolic conversion are generally electrophilic in nature and are either cations or neutral compounds, including reactive oxygen species and free radicals.

Metabolic reactive intermediate-mediated oxidative stress can cause damages to cellular components such as impairment of protein functions, membrane lipid peroxidation, and formation of DNA adduct. Oxidative stress plays a crucial role in the pathogenesis processes that attribute to various disease conditions. Detoxification enzymes and antioxidants act as defense mechanisms against metabolic reactive intermediates or metabolites. Nuclear factor erythroid-2-related factor 2 (Nrf2)–antioxidant response element pathway is critical to the mechanism of action of detoxification enzymes.

Important factors in contributing to individual variations in response to xenobiotics-mediated toxicity and carcinogenesis include genetic polymorphisms of metabolic enzymes, environmental factor, and lifestyle. Extensive investigations in the past decades have discovered a variety of chemical compounds that are capable of modulating metabolic enzymes for health benefits. Many of them are present in diets, such as sulfur- and non-sulfur-containing compounds, organosulfur compounds, flavonoids and carotenoids in vegetables, polyphenols and triterpenes in fruits, polyphenols in green tea, and unsaturated fatty acids in algae.

This book addresses the above subjects in broad scope with emphasis on biochemical and biomedical aspects and their recent advances. It can be used as supplemental text materials for senior undergraduates and graduate students in the areas of biochemistry, biomedical sciences, pharmacology, and environmental sciences. Researchers may also be interested in this book to gain knowledge on foreign compound metabolic enzymes and their health implications. There are 19 chapters in this book, including those that are extensively updated and revised from my 2012 Springer book. A bibliography is provided in each chapter, where references are cited for readers who wish to pursue topics in more detail.

The author wishes to thank Professor David Carpenter, Director of the Institute of Health and the Environment, University at Albany, for his encouragement and suggestions, and Professor Norman L. Strominger of Albany Medical College for his suggestions and comments. The author also wishes to express appreciation to the editorial board at Springer Nature for their assistance.

Rensselaer, NY, USA Chang-Hwei Chen

Contents

Chapter 1
Overview

Xenobiotics are substances that are foreign to an organism and are not produced by a living organism. The Greek word "xeno" in the word "xenobiotics" means foreigner, and the word "biotic" refers to belonging to life or relating to living things. Humans are constantly exposed to foreign compounds, including foods, drugs, other chemicals, and cigarette smoke. Foods are consumed to supply essential nutrition and to provide energy to maintain life. Drugs are chemical substances used in treatment, diagnosis, or prevention of diseases. Antibiotics and hormones are applied to raise chicken, cattle, and pig. Other chemicals such as fertilizers are utilized to grow agriculture products. Preservatives are used to food industries. Organic solvents are applied in household as cleaning products. Pollutants including smokes are released into air from industrial factories.

Some of the foreign compounds are water soluble (hydrophilic), while a wide range of them are not soluble in water (lipophilic). After their digestion and inhalation in the body, metabolism of foreign compounds is a major challenge. This challenge is achieved by low specificity metabolic enzymatic systems, which make it possible to metabolize a wide range of foreign compounds, either hydrophilic or lipophilic. Metabolic enzyme-catalyzed reactions are essential in facilitating the elimination of foreign compounds from the body.

Metabolic enzymes produced from information stored in the genes play central roles in biotransformation and detoxification of foreign compounds. Before they are excreted from the body, foreign compounds undergo two distinctive metabolic processes, which are generally classified as activation metabolism (phase I) and detoxification metabolism (phase II). Lipophilic compounds need to go through both activation and detoxification processes, while hydrophilic compounds can bypass the step of activation.

Some foreign compounds are acutely or potentially toxic, while others initially exhibit no intrinsic toxicity but become harmful after metabolic conversion. In many cases, metabolic conversion catalyzed by activation enzymes produces highly reactive metabolites or metabolites that may exhibit toxicity. Meanwhile, detoxification

C.-H. Chen, *Xenobiotic Metabolic Enzymes: Bioactivation and Antioxidant Defense*,
https://doi.org/10.1007/978-3-030-41679-9_1

enzyme-catalyzed reactions detoxify and defend the body against potential harmful reactive intermediates and metabolites, including reactive oxygen species and free radicals.

Reactive intermediates are electron-deficient species or positively charged ions, and free radicals are atomic or molecular species with unpaired electrons. These electrophilic reactive intermediates are able to react with neutrophilic donors from target cellular molecules such as proteins, lipids, and DNA. Oxidative stress occurs as the production of reactive intermediates is not balanced by the capability of the body's defense systems, leading to implicating the pathological conditions in association with inflammation and carcinogenesis. Uncontrolled inflammatory response can result in cell and tissue damages, which play a crucial role in the pathogenesis processes that attribute to cardiovascular disorder, neurodegeneration, pulmonary injury, and cancers.

Nuclear factor erythroid-2-related factor 2 (Nrf2)-antioxidant response element (ARE) pathway is critical to the elimination of reactive chemical intermediates. Its activation plays an important role in the defense mechanism against electrophile-mediated oxidative damages. Nrf2 is a transcription factor that responds to oxidative stress by binding to ARE in the promoter of genes coding for antioxidant proteins and detoxification enzymes. As the body's defense mechanism, Nrf2-ARE pathway acts as a sensor to oxidative stress through stimulation of antioxidant proteins and detoxification enzyme genes.

Genetic variations in metabolic enzyme expression are an important feature associated with foreign compound-mediated biomedical and biochemical effects. Genetic polymorphisms have a significant effect on an individual's susceptibility to diverse toxicity. The differences in individual sensitivity to foreign compound metabolites could explain many observed differential reactions to foreign compounds among individuals.

Enzyme inducibility also plays a role in the susceptibility of individuals to foreign compound-mediated toxicity. An increase in detoxification enzyme activities that detoxify electrophilic reactive intermediates can intervene in foreign compound-mediated biomedical or biochemical effects. The consequence of detoxification enzyme induction also depends on the balance between the rates of activation and detoxification expressions.

Extensive research has been carried out to investigate compounds that modulate the activity of metabolic enzymes. A diversity of metabolic enzyme inducers have been reported.

Phytochemicals present in human daily diets include sulfur- and non-sulfur-containing compounds in cruciferous vegetables, thiols and sulfides in allium vegetables, flavonoids, carotenoids and curcumin in root vegetables, and polyphenols and triterpenes in fruits.

1.1 Xenobiotics That Humans Are Exposed To

Humans are constantly exposed to foreign compounds, including foods, drugs, and household and environmental chemicals, which are either naturally produced or artificially manufactured, but not naturally present in living cells.

1.1.1 Foods and Drugs

Metabolites and wastes are produced when foods are converted into raw materials and energy through biochemical reactions occurring in living cells. Medicine used to fight diseases can accumulate in the body and cause unwanted side effects. In some cases, metabolisms of foods and drugs produce reactive intermediates and metabolites, which can interact with cell components and cause cellular and functional damages. Allelic variants of some drug-metabolizing enzymes allow to determine individuals with the susceptible genotypes. Genetic polymorphisms of drug-metabolizing enzymes are associated with toxicity or therapeutic failure of pharmacotherapy.

1.1.2 Smoke and Household and Environmental Chemicals

Cigarette smoke contains various toxic chemicals including nitrosamines, nicotine, benzo[a]pyrene, and other polycyclic aromatic hydrocarbons. The association of cigarette smoke with chronic inflammation and disease conditions is well documented. Organic solvents such as benzene and phenol are present in household products including detergents, paints, and pesticides. Organic solvents are also applied to synthesize resins, nylon, synthetic fibers, and plastics in manufacture industries. Petroleum refinery, incineration, diesel exhausts, and cool plants produce industrial pollution including polycyclic aromatic hydrocarbons, dioxins, and polychlorinated biphenyls (PCB). Arsenic present in polluted drinking water is a risk factor for cancer.

1.2 Metabolisms of Foreign Compounds Before Excretion

Based on the solubility, foreign compounds that find their way into the body are either water soluble (hydrophilic) or fat soluble (lipophilic). While many foreign compounds are water soluble, a large number of them are not. Before being eliminated from the body, lipophilic compounds need to be converted into hydrophilic compounds. To carry out such conversion, the body develops metabolic enzyme systems.

Metabolic enzymes are produced from information stored in the genes. There are two distinctive metabolic processes, which are generally classified as phase I and phase II metabolisms, and their respective enzymes are called activation enzymes and detoxification enzymes.

1.2.1 Activation Enzymes

Action enzymes catalyze functionalization reactions, in which a functional group is introduced to the chemical structure of a foreign compound through oxidation, reduction, or hydrolysis reaction. Functionalization increases the polarity of a lipophilic foreign compound, making it more soluble in water. In many cases, the modified compound leads to the formation of reactive intermediate or metabolite. However, a hydrophilic foreign compound that has a functional polar group may bypass this functionalization process.

1.2.2 Detoxification Enzymes

The functionalized foreign compound is then further metabolized and catalyzed by detoxification enzymes. Detoxification metabolism is generally involved with conjugation reactions, where a functionalized foreign compound is combined with a small molecule (referred to as cofactor) to form a less active conjugate. The conjugation reaction generally further increases the solubility and the excretory potential of a foreign compound, thus facilitating its removal from the body.

1.2.3 Catalytic Reactions: Functionalization and Conjugation

Activation enzymes catalyze a variety of reactions that introduce functional groups to foreign compounds, where the reactions take place at specific atoms or groups. Detoxification enzymes catalyze various conjugation reactions to form conjugated compounds that detoxify and facilitate the excretion of foreign compounds. Either activation or detoxification metabolism consists of a variety of enzymes that carry out different catalytic reactions. Knowledge about these catalytic reactions and their functional characteristics is fundamental to understanding how metabolic enzymes act on foreign compounds and how foreign compounds are metabolized by certain enzymes.

1.3 Potential Toxicity of Metabolic Reactive Intermediates

Foreign compound metabolisms are beneficial to living organisms. However, in many cases, the generated reactive intermediate or metabolite is potentially toxic. Moreover, a significant number of reactive intermediates have potential to react with oxygen to form reactive oxygen species including free radicals, which can further interact with cellular components. The toxicity caused by the accumulation of reactive intermediates in the body could contribute to a variety of disease conditions, such as aging, cardiovascular disease, neurological disorders, and cancer.

1.3.1 Electrophiles

Reactive intermediates produced by functionalization reactions are generally electrophilic in nature. An electrophile is an electron acceptor, a cation, or a neutral compound, but can never be an anion. Reactive intermediates can further interact with oxygen molecules to produce reactive oxygen species such as highly reactive cationic electrophiles and free radicals. Reactive electrophilic intermediates are capable of covalent bonding with nucleophilic sites of cellular molecules to form stable adducts with target proteins, DNA, or lipids.

1.3.2 Interactions with Cellular Components

Reactive intermediates generated by metabolism of foreign compounds have the capacity to interact with cellular components, including proteins, DNA, and lipids, ultimately leading to toxic effects. The interactions of reactive intermediates with cellular components can be either specific or non-specific. Specific interaction involves binding to a specific site in a receptor, enzyme, or transport protein, which may trigger alteration in the structure and function of target molecules.

1.3.3 Foreign Compound-Mediated Oxidative Stress

Oxidative stress occurs as the production of reactive intermediates and metabolites is not balanced by the capability of the body's defense systems to detoxify them. Oxidative stress-mediated cellular damages play a crucial role in the pathogenesis processes that can attribute to inflammation and disease conditions.

1.3.4 Cellular Functional Damages

Interactions of reactive intermediates with cellular proteins lead to oxidative damage. If the oxidative damage is not repaired, loss of catalytic function for enzyme and impairment of function for protein can occur. Membrane lipid peroxidation, which is referred to as the oxidative degradation of membrane lipids, has been implicated as one of the mechanisms that cause xenobiotics-induced cellular functional damages. Reactive chemical intermediates are also capable of causing DNA damage, which eventually may attribute to cell transformation and tumor initiation. The oxidation of a DNA base has severe consequences on DNA base pairing.

1.4 Genetic and Environment Factors Affecting Xenobiotic Metabolism

Foreign compound metabolisms exhibit a great deal of individual variability and are affected by genetics. Lifestyle and environmental factors can also play a role. Chronic alcohol consumption and cigarette smoking also influence xenobiotic metabolism by affecting metabolic activation.

1.4.1 Genetic Polymorphisms

Metabolic enzyme expressions may vary among individuals due to genetic polymorphisms. Variations in genetic polymorphisms can affect an individual's susceptibility to foreign compound-mediated toxic effects. An unusually high expression of activation enzyme may give rise to an overload of reactive intermediates or metabolites, while an extraordinary low efficiency of detoxification enzyme may result in inefficient detoxifying metabolism. Maintaining reactive intermediates and metabolites at a minimum level depends on the relative efficacies of activation enzymes and detoxification enzymes. A delicate balance between the reactions catalyzed by these two classes of enzymes is essential.

1.4.2 Environment and Lifestyle

The expressions of metabolic enzymes may vary among individuals due to environmental factor and lifestyle. Exposure to environmental chemicals and chronic consumption of alcohol or cigarette smoke are capable of modulating the expression of xenobiotic metabolizing enzymes.

1.5 Nrf2-ARE Pathway: Defense Mechanism

Nrf2-ARE pathway is critical to the mechanism of action of metabolic enzymes. In the liver, Nrf2 regulates cytoprotective genes, which confer cellular resistance to toxicity and oxidative stress, thus providing protection against inflammation and various chronic diseases. As the body's defense mechanism, Nrf2-ARE pathway acts as a sensor to oxidative stress through stimulation of antioxidant proteins and detoxification enzyme genes. Oxidative stress and inflammation are evidenced as a pathological mechanism, leading to various disease conditions.

1.6 Defenses Against Oxidative Stress Mediated by Foreign Compounds

The body's defense system against oxidative stress includes endogenous antioxidants and exogenous antioxidants. Endogenous antioxidants are products of the body's metabolism, which can be either enzymatic or nonenzymatic. Metabolic detoxifying enzymes are endogenous enzymes, which are present as a cell's defense mechanism against oxidative stress-mediated damages.

1.6.1 Antioxidants

Important enzymatic antioxidants include catalase, glutathione peroxidase, and glutathione reductase. Nonenzymatic endogenous antioxidants are glutathione, uric acid, melatonin, and bilirubin, which are molecules characterized by their ability to inactivate radicals and oxidants. Exogenous antioxidants are available as dietary supplements such as vitamins, lutein, beta-carotene, and lycopene.

1.6.2 Modulation of Metabolic Enzymes for Health Benefits

The modulation of metabolic enzymes to reduce foreign compound-mediated toxic effects has been a subject of interest in the past decades. To protect individuals against foreign compound-mediated toxicity, important hypotheses have been postulated. An increase in dietary inducers of detoxification enzymes is considered as a promising proposal to minimize foreign compound-mediated toxic effects.

1.6.3 *Inducibility of Metabolic Enzymes*

An important toxicologically relevant feature associated with metabolic enzymes is the potential of these enzymes for induction or inhibition by some chemical compounds. The inducibility of these enzymes makes it possible to modify their expression for health benefits.

1.7 Dietary Inducers of Detoxification Enzymes

Many chemical compounds that are capable of acting as enzyme modulators are present in the daily human diet. Extensive research has been carried out to explore how these enzymes can be modulated in human diets for health benefits. Cruciferous vegetables contain inducers of detoxification enzymes including sulfur-containing compounds (isothiocyanates, phenethyl isothiocyanate, sulforaphane, and 1,2-dithiole-3-thione) and non-sulfur-containing compound (indole-3-carbinol).

Enzyme inducers such as organosulfur compounds (thiols and diallyl sulfides) are present in allium vegetables, and flavonoids, carotenoids, and curcumin are present in root vegetables. Fruits consist of detoxification enzyme inducers including polyphenols (quercetin, kaempferol, pomegranate, anthocyanin, and procyanidin) and triterpenes (limonoids and olive oil).

1.7.1 *Vegetables and Fruits*

Diets rich in vegetables and fruits that contain enzyme modulators have received much attention, in particular, those rich in inducers of detoxification enzymes, such as cruciferous, allium, and root vegetables. Inducer compounds include sulfur-containing, non-sulfur-containing, and organosulfur compounds, as well as polyphenols, flavonoids, and carotenoids.

1.7.2 *Green Tea and Algae*

Green tea polyphenols, especially epigallocatechin-3-gallate, were found to modulate metabolic detoxification enzymes. Green tea polyphenols were also shown to exert cancer-protective activity by inhibiting the metabolic activation of carcinogens, particularly by CYP450. Methanolic extract of seaweed was found to revert dioxin-induced toxicity through an alteration in the expression of antioxidants. Polysaccharide extract from brown alga was reported to inhibit the carcinogen activator CYP450 and induce the carcinogen detoxification enzymes glutathione *S*-transferases.

Bibliography

Anzenbacher A, Zanger UM (eds) (2012) Mechanism of drugs and other xenobiotics. Wiley-VCH, Hoboken
Boelsterli UA (2007) Mechanistic toxicology. CRC Press, Boca Raton
Buxton ILO (2006) Pharmacokinetics and pharmacodynamics. In: Brunton LL et al (eds) Goodman & Gilman's the pharmacological basis of therapeutics. McGraw-Hill, New York
Chen C-H (2012) Activation and detoxification enzymes: functions and implications. Springer, New York
Conney AH (2003) Enzyme induction and dietary chemicals as approaches to cancer chemoprevention. Cancer Res 63:7005–7031
Finley JW, Schwass DE (1985) Xenobiotic metabolism: nutritional effects. American Chemical Society, Washington, DC
Giacomini KM, Sugiyama Y (2006) Membrane transports and drug response. In: Brunton LL et al (eds) Goodman & Gilman's the pharmacological basis of therapeutics. McGraw-Hill, New York
Gonzalez FJ, Tukey RH (2006) Drug metabolism. In: Brunton LL et al (eds) Goodman & Gilman's the pharmacological basis of therapeutics. McGraw-Hill, New York
Hodgson E, Das PC, Cho TM, Rose RL (2008) Phase I metabolism of toxicants and metabolic interaction. In: Smart RC, Hodgson E (eds) Molecular and biochemical toxicology. Wiley, New York
Ioannides C (2002) Enzyme systems that metabolise drugs and other xenobiotics. Wiley, New York
Jakoby WB, Bend JR, Caldwell J (1982) Metabolic basis of detoxication. Academic Press, New York
Jakoby WB (1980) Enzymatic basis of detoxication, vol 2. Academic, New York
Josephy PD, Mannervik B, de Montellano PO (1997) Molecular toxicology. Oxford University, New York
LeBlanc GA (2008) Phase II - conjugation of toxicants. In: Smart RC, Hodgson E (eds) Molecular and biochemical toxicology. Wiley, New York
Lee JS, Obach RS (2003) Drug metabolizing enzymes. Marcel Dekker, New York
Mulder GJ (1990) Conjugation reactions in drug metabolism: an integrated approach. Taylor and Francis, London
Parkinson A, Ogilvie BW (2008) Biotransformation of xenobiotics. In: Klaassen CD (ed) Casarett & Doull's toxicology: the basic science of poison. MacGraw-Hill, New York
Sardesai VM (2003) Introduction to clinical nutrition. Marcel Dekker, New York
Smart RC, Hodgson E (2008) Molecular and biochemical toxicology. Wiley, New York

Chapter 2
Foreign Compounds: Foods, Drugs, and Other Chemicals

Foods are substances that contain essential nutrients such as proteins, carbohydrates, and minerals to provide energy and maintain life. Drugs are chemical substances for a living organism to produce a biological effect and affect how the body works. A medicine is a drug that is used in treatment, diagnosis, or prevention of a disease. Chemicals are compounds that are artificially made or occur naturally. They are a major part of the pharmaceutical industry.

Chemicals are also used for preservation in food industry. Fertilizer chemicals are applied to accelerate the growth of crops. Chemical industries produce plastics, rubber, and fibers as household products. In producing such products, chemical factories release millions of tons of toxic chemicals and organic solvents into the atmosphere.

Xenobiotics include foods, drugs, cigarette smoke, household products, and environmental chemicals. Table 2.1 presents foreign compounds that humans are exposed through digestion or inhalation.The potential health concerns associated with foreign compound metabolisms are also discussed in this chapter.

2.1 Foods

Vegetables and fruits that humans ingest are largely natural plants. Natural plants generate a variety of biologically active chemicals to protect themselves. A diverse group of chemicals are present in various plants used as food such as grains and fruits. After ingestion, metabolites are generated via complex biochemical reactions that occur continuously within the cells. Contaminants in food occur, including pesticides from crop sprays, fungi from storage, phthalate ester from packaging, and styrene from containers. Moreover, residues of antibiotics and hormones used to raise chickens, cattle, pigs, and sheep may remain as contaminants in meat. Fish may be contaminated with industrial wastes such as mercury, PCBs, and dioxins.

C.-H. Chen, *Xenobiotic Metabolic Enzymes: Bioactivation and Antioxidant Defense*,
https://doi.org/10.1007/978-3-030-41679-9_2

Table 2.1 Typical foreign compounds that humans are exposed to

Foods	Drugs	Cigarette smoke	Household products	Environmental chemicals
Heterocyclic amines	Acetaminophen	Nitrosamines	Benzyne	Diesel exhausts
Nitrosamines	Xanthine	Nicotine	Phenol	Arsenic
Polycyclic aromatic hydrocarbons	Terfenadine	Benzo[a]pyrene		Polychlorinated biphenyl
	Menadione			
Azo dyes	Diazepam			Dioxins
α,β-unsaturated aldehydes				
Mycotoxin				

Molds may produce metabolites with the potential to produce adverse health effects. Chemical derivatives are produced, when meat or fish is cooked at high temperature. Known potentially toxic chemicals in foods include heterocyclic amines, nitrosamines, polycyclic aromatic hydrocarbons, azo dyes, unsaturated aldehydes, and mycotoxins. These chemicals are discussed below.

2.1.1 Heterocyclic Amines

Heterocyclic amines are formed when amino acids react with creatine in muscle when meats (beef, pork, or fish) are cooked at high temperature by frying, broiling, or barbecuing.

Heterocyclic amines are organic compounds that contain at least one atom of carbon and at least one atom of nitrogen within aromatic or nonaromatic rings such as pyridine or pyrimidine. More than a dozen heterocyclic amines have been identified. 2-Amino-1-methyl-6-phenylimidazo [4,5-beta] pyridine is the most abundant heterocyclic amine found in human diets.

Heterocyclic amines are enzymatically activated by activation enzymes such as cytochrome P450. Many heterocyclic amines have been reported to be carcinogenic. Chronic administration of these chemicals was found to induce tumors in rats at several sites including the colon. Green tea and black tea that inhibit activation enzymes were reported to impede heterocyclic amine-induced mutagenesis.

2.1.2 Nitrosamines

Nitrosamines are found in trace amounts in foods such as mushrooms, fermented and smoked fish, and pickled foods. Cured meats such as bacon can also contain nitrosamines because sodium nitrite is added as a preservative. High cooking temperatures used to fry bacon also contribute to nitrosamine formation.

Nitrosamines have also been found in human gastric juice, possibly formed by reaction between amines and nitrites from the diet.

Nitrosamines containing an organic functional group N–N=O are formed by the addition of N=O group to secondary or tertiary amines. N-nitroso derivatives of amines are formed by reaction between nitrite and amines. Chemical carcinogens generally require metabolic activation and then to bind to DNA, leading to mutation and develop tumors. Humans seem to be susceptible to carcinogenic properties of nitrosamines, and high temperature cooking may be responsible for high cases of colon cancer.

Human susceptibility to nitrosamine toxicity varies from individual to individual, depending on the expression of metabolic enzymes. Nitrosonornicotine was found to significantly increase activation enzyme expression in the liver and the lung and to noticeably decrease the expression of glutathione levels and glutathione *S*-transferase in rats.

2.1.3 Polycyclic Aromatic Hydrocarbons

Humans could be exposed to polycyclic aromatic hydrocarbons by eating grilled, charred meats. Polycyclic aromatic hydrocarbons are products of incomplete combustion at high temperatures. Major polycyclic aromatic hydrocarbons detected in charcoal meats include benzo[a]pyrene and dibenz[*a,h*]anthracene. Regular consumption of overcooked charcoal barbecued beef has also been reported to associate with increased levels of colon cancer.

In addition to their present in grilled or charred meat, polycyclic aromatic hydrocarbons also exist in cigarette smoke and auto exhausts due to the combustion of organic materials. Polycyclic aromatic hydrocarbons exhibit their toxic properties after metabolic conversion to chemically active intermediates. Two major carcinogens in cigarette smoke are benzo[a]pyrene and dibenz[a,h]anthracene.

Following absorption, dibenz[*a,h*]anthracene is distributed to various tissues with the highest accumulation in the liver and kidneys. There is evidence that links polycyclic aromatic hydrocarbons to the induction of activation enzymes. Polycyclic aromatic hydrocarbons are metabolically activated by activation enzymes such as cytochrome P450 (CYP450). These reactive species have the capacity of interacting with DNA.

A considerable number of studies over the years have documented the link between benzo[*a*]pyrene and cancers. Benzo[a]pyrene is a major lung carcinogen, which toxicity is produced by the bioactivation of benzo[a]pyrene to a toxic intermediate, benzo[a]pyrene-7,8-dihydrodiol-9,10-epoxide. Dibenz[a,h]anthracene is metabolized mainly by CYP450 to form carcinogenic metabolites. Animal studies reported that dibenz[*a,h*]anthracene depresses immune responses in mice. Dibenz[*a, h*]anthracene was also shown to induce hepatic aryl hydrocarbon hydroxylase activity in mice.

2.1.4 *Azo Dyes*

Azo dyes are stable upon heating and are also stable in the pH ranges of food products, and their colors do not fade when exposed to light or oxygen. These properties make azo dyes applicable for use in foods. However, the azo linkage is the most labile portion of the azo dye molecule. Azo dyes undergo enzymatic breakdown catalyzed by azo-reductase presence in various microorganisms.

Azo-reductase activity is high in the liver and the kidney. After cleavage of the azo-linkage, amine metabolites are absorbed in the intestine and excreted in the urine. Many azo pigments are nontoxic. The acute toxicity is low in consuming azo dye-colored foods. However some azo dyes have been banned for food use because of side effects due to their degradation products.

Azo degradation products have been reported to be mutagenic or carcinogenic. A typical example of mutagenic pigment is orthonitroaniline orange. Enzymatic sulfonation appears to decrease the toxicity of azo degradation products by facilitating urinary excretion of azo dyes and their metabolites.

2.1.5 *α,β-Unsaturated Aldehydes*

Cooking oils that are used repeatedly in caterings and households can generate significant amounts of oxygenated α,β-unsaturated aldehydes. 4-Hydroxynonenal is an α,β-unsaturated hydroxyalkenal generated in the oxidation of lipids or corresponding fatty acids. It is present in higher quantities as the lipid peroxidation chain reaction increases. Through diets, the body absorbs 4-hydroxynonenal and oxygenated α,β-unsaturated aldehydes. These compounds have received a great deal of attention because they are considered as potentially contributing agents to a number of disease conditions such as inflammation, respiratory distress syndrome, and diabetes.

2.1.6 *Mycotoxin*

Mycotoxin is a toxin produced by fungi of mushrooms, molds, or yeasts. When fungi propagate into colonies, mycotoxin levels become high. Wild mushrooms contain an assortment of mycotoxins that can cause noteworthy health problems. Mycotoxin appears in food chains as a result of fungal infection of crops and can also remain in food chains of meat and dairy products.

Patulin, a mycotoxin produced by a variety of molds, is commonly found in rotting apples. Some studies reported that patulin is toxic and may be a carcinogen. Studies of rat liver tissues reported that patulin decreases glutathione *S*-transferase activity and markedly increases lipid peroxidation. These effects may be a result of patulin-mediated reduction in the level of glutathione. The decrease in glutathione level and glutathione *S*-transferase activity may be related to the presumed mutagenic or carcinogenic potential of patulin.

Poultry such as turkey is quite susceptible to toxic effects of aflatoxin B_1, another mycotoxin, due to a combined result of efficient activation by CYP450 and deficient detoxification by glutathione *S*-transferases. Aflatoxin B_1 can also cause health problems by acting as an allergen or irritant or by weakening the immune system.

2.2 Drugs

The majority of medications that humans take are synthetic chemical compounds. Individuals may experience some side effects on taking specific over-the-counter or prescription drugs. Such side effects derived from drug metabolisms make the accumulation of drugs in the body harmful. This can occur with drugs such as pain killers and diuretic. Acetaminophen is a major ingredient in pain killers. An overdose of acetaminophen can cause damage to the liver. Furosemide, a diuretic, if given in excessive amounts, can lead to profound diuresis with depletion of water and electrolytes.

Active ingredients of drugs perform the needed chemical or biochemical actions, leading to therapeutic responses in the body. The time course and the duration of action are affected by the activities of enzymes associated with the metabolisms of these drugs. When the active constituent of a drug is metabolized to a chemically active intermediate or metabolite, drug-mediated toxic effects may occur if detoxification enzymes are inefficient in eliminating it from the body.

Overdoses of drugs are a health concern. Harmful effects can also be attributed to the accumulation of drug ingredients in the body. Examples of potential overdose of drugs include acetaminophen, xanthine, terfenadine, menadione, and diazepam that are described below. A low expression of detoxification enzymes corresponding to the metabolism of these drugs is a major concern.

2.2.1 Acetaminophen

Acetaminophen is one of the most common pharmaceutical agents that are involved in overdose toxicity. The hepatotoxicity associated with misuse or overdose of acetaminophen is well documented. This drug is metabolically activated via CYP450 system, while glucuronide and sulfonate conjugates are involved in hepatic metabolism of acetaminophen elimination. In the liver, acetaminophen is metabolized by conjugation reaction to form water-soluble conjugate ready for elimination in the urine. However, in the case of a deficiency in conjugation enzyme UDP-glucuronosyl transferase, the accumulated reactive intermediate or metabolite is capable of interacting with cellular proteins and hepatocyte membranes, resulting in hepatocellular damages.

2.2.2 *Xanthine*

Bronchodilators xanthine (3,7-dihydro-purine-2,6-dione) and its derivatives are a group of alkaloids that are commonly used as mild stimulants and in treatment of the symptoms of asthma. Xanthine is a product of the pathway of purine degradation. Metabolic conversion of xanthine to uric acid, carried out by an enzymatic reaction catalyzed by xanthine oxidase, leads to subsequent excretion of xanthine from the body.

In the case of a lack of sufficient xanthine oxidase, xanthine cannot be readily converted to uric acid, and an accumulation of xanthine in the body occurs, which could result in xanthine-mediated oxidative stress. In cardiomyocyte studies, pretreatment of detoxification enzyme inducers, such as resveratrol or 1,2-dithiole-3-thione, was found to increase resistance to xanthine-mediated effects.

2.2.3 *Terfenadine*

Terfenadine, an antihistamine, is metabolized by CYP450 isozyme to its metabolite fexofenadine. This metabolism is blocked by drug such as erythromycin or food such as grapefruit, making it difficult to metabolize and remove terfenadine from the body. An elevated level of terfenadine can lead to an adverse cardiac effect on the heart rhythm, although its metabolite may not have such an effect. Potential toxicity can also occur as a result of its interaction with other medications. This drug was removed from the market in 1997.

2.2.4 *Menadione*

Menadione (vitamin K3) is one of man-made versions of vitamin K. Menadione is a lipophilic vitamin precursor that is converted into menaquinone in the liver. A deficiency in vitamin K, especially in infants can be a severe health issue, because of easily suffering from extensive hemorrhaging, while an overdose of menadione can be as equally detrimental. Overdoses of menadione have been reported to cause adverse outcomes including hemolytic anemia, neonatal brain, or liver damage. Menadione supplements are therefore banned because of their potential toxicity in human use.

2.2.5 *Diazepam*

Diazepam belongs to a class of drugs known as benzodiazepines. It is used to treat anxiety, alcohol withdrawal, and seizures. Studies were conducted to ascertain any involvement of free radical-mediated prooxidative processes in different brain regions following diazepam administration. A significant decrease in thiobarbituric acid reactive substance formation was observed in cerebral cortex. Glutathione reductase activity was also decreased in cerebellum and brain stem.

2.3 Smoke

Humans are exposed to cigarette smoke and auto exhausts more commonly than other environmental sources. Cigarette smoke contains a great variety of chemicals many of which are toxic. Among them are nitrosamines and polycyclic aromatic hydrocarbons. The association of cigarette smoke with a higher level of chronic inflammation and other disease conditions is well documented. Typical toxic chemicals produced in cigarette smoke include nitrosamines, nicotine, and benzo [a]pyrene.

2.3.1 *Nitrosamines*

There are tobacco-specific nitrosamines that are found only in tobacco products and are generated during fermentation, curing, and burning of the tobacco leaf. Tobacco-specific nitrosamines are strongly carcinogenic in laboratory animals. Many nitrosamines have been reported to be carcinogenic in a wide variety of experimental animals. For example, *N*-nitrosodibutylamine and its hydroxylated metabolite (*N*-nitrosobutyl(4-hydroxybutyl)amine) are urinary bladder-specific carcinogens. Considerable evidence supports the role of tobacco-specific nitrosamines as an important contributing factor for cancers of the lung, pancreas, and esophagus in humans.

2.3.2 *Nicotine*

Cigarette-generated nitrosamines are composed of various amines such as nicotine and nornicotine. Nicotine in cigarette smoke is rapidly absorbed into the bloodstream and reaches the brain within 10 s. Nicotine affects the brain chemistry, resulting in a number of chemical reactions that involve hormones and neurotransmitters.

2.3.3 *Benzo[a]pyrene*

Humans are exposed to cigarette smoke and auto exhausts more commonly than other environmental sources. Benzo[a]pyrene is found in cigarette smoke resulting from the combustion of organic materials. Benzo[a]pyrene was found to have the capacity of binding to cell components in major organs.

2.4 Household products

Because of their volatility, humans are exposed to many organic solvents through inhalation. Organic solvents are present in common household and industrial products such as gasoline, paint remover, varnishes, and wood sealants. Household products including lubricants, detergents, paints, and pesticides also contain a variety of organic solvents. Organic solvents are also used to make resins, nylon, synthetic fibers, and plastics in industries. Volatile organic solvents are also present in dry cleaning shops, electronics industries, and scientific laboratories. Benzene and phenol are typical examples of organic solvent present in household products.

2.4.1 *Benzene*

Exposure to benzene is known to cause human health problems. The toxic effects of benzene in humans are attributed to hydroxylated metabolites such as hydroquinone and phenol.

Benzene is present in unleaded gasoline and cigarette smoke. Long-term exposure of benzene can have harmful effects on the bone marrow and can cause a decrease in red blood cells, leading to anemia. Benzene was also found to affect the immune system, causing an increase in the chance for infection, and to depress the immune system and cause cancer in both animals and humans.

2.4.2 *Phenol*

Phenol is the primary metabolite of benzene. Detoxification enzyme-catalyzed conjugation reactions are involved in detoxifying phenol. At a low exposure concentration of benzene, phenylsulfonate is the major conjugate of phenol in the blood. However, at a high exposure concentration, phenylglucuronide is the predominant conjugate. Reductions in spleen weight and white blood cell numbers were reported to correlate with the concentration of phenylsulfonate in the blood.

2.5 Environmental Chemicals

Industrial combustions in refinery, incineration, and coal plants produce industrial pollution including polycyclic aromatic hydrocarbons, dioxins, and polychlorinated biphenyls (PCBs). Transportation vehicles generate exhaust gases that also consist of toxic substances. Known environmental toxicants that humans are exposed to include diesel exhaust particles generated from transportation vehicle operation and arsenic and polychlorinated biphenyls present in contaminant drinking water.

Environmental toxicants to which humans are exposed are of major concern. Potentially harmful compounds are either intrinsically toxic or become toxic after metabolic conversion. They have been reported to play an important role in the pathogenesis of lung disease and other disease conditions such as cancer.

2.5.1 Diesel Exhausts

Diesel exhausts produced in combustion of diesel fuel are a mixture of gases and fine particles that contain harmful air contaminants such as polycyclic aromatic hydrocarbons. Diesel exhausts also include many potential cancer-causing substances such as benzene, arsenic, and formaldehyde, as well as other toxic pollutants such as nitrogen oxides. Diesel exhaust particles can initiate and exacerbate airway allergic responses through enhanced IgE production.

Exposure to diesel exhausts can cause inflammation in the lungs, aggravating chronic respiratory symptoms and increase the frequency or intensity of asthma attacks. Such harmful effects are especially pronounced in individuals whose detoxification enzyme expression is impaired. Chemically reactive intermediates are believed to play a key role in cellular damage after exposure to diesel exhaust particles. Glutathione *S*-transferase was reported to involve in the detoxification of diesel exhaust particle-mediated allergic inflammation.

2.5.2 Arsenic

Besides organic solvents, humans are also exposed to nonvolatile chemicals such as arsenic compounds and polychlorinated biphenyls. Arsenic, a risk factor for cancer, is present in polluted drinking water. Populations in many regions of the world are chronically exposed to arsenic-contaminated drinking water. When rice is grown in polluted water, paddy rice takes up arsenite readily from soaking soil.

Arsenic is metabolized to monomethylarsonic acid, which is converted to dimethyarsinic acid by methyltransferase enzymes. The metabolite requires *S*-adenosyl-methionine as the methyl donating cofactor before its excretion through urine. A remarkable diversity in arsenic methyltransferase activity may account for the wide variability in the susceptibility of individuals to arsenic toxicity.

Ground water pollution by arsenic is a serious worldwide problem. Drinking water from arsenic-tainted wells causes ailments marked by rough skin and often leads to serious diseases such as skin or bladder cancer. Arsenic was also reported to interfere with methyltransferases and inactivate tumor suppressor genes. Other studies reported that arsenic-induced malignant transformations are linked to DNA hypomethylation.

2.5.3 Polychlorinated Biphenyls

Polychlorinated biphenyls (PCBs), a major class of persistent organic pollutants, are a group of man-made organic chemicals that consist of carbon, hydrogen, and chlorine atoms. Due to non-flammability and chemical stability, PCBs are used in a variety of industrial and commercial applications. PCBs do not readily break down and can remain for long periods in water and soil.

PCBs are metabolized to hydroxylated compounds. Many of these metabolites are further converted to either the sulfonate or the glucuronic acid conjugates catalyzed by detoxification enzymes, such as sulfotransferases or uridine diphosphate glucuronosyl transferase. The efficiency of glucuronidation depends on the structure of the PCB metabolites. PCBs were also reported to induce the activity of monooxygenases that catalyze the metabolism of PCBs, leading to formation of metabolites and potential adverse health effects.

Studies in humans support evidences for potential carcinogenic and non-carcinogenic effects of PCBs. They have been shown to cause cancer in animals as well as a number of serious non-cancer health effects in animals, including the effects on the immune, reproductive, nervous, and endocrine systems.

2.5.4 Dioxins

Dioxins are environmental pollutants. Its chemical name is 2,3,7,8-tetrachlorodibenzo-para-dioxin (TCDD). Dioxins are mainly by-products of industrial processes, including smelting, chlorine bleaching of paper pulp, and the manufacturing of herbicides and pesticides. Dioxins can also result from natural processes such as volcanic eruptions and forest fires.

Dioxins exert their effects through interaction with a specific intracellular protein, the Ah receptor. The Ah receptor functions as a transcriptional enhancer, which interacts with a number of other regulatory proteins. Dioxins were also reported to modulate interactions with specific base sequences in the DNA.

Short-term exposure to high levels of dioxins may result in skin lesions and liver function alteration. Long-term exposure is linked to impairments of immune, nervous, and endocrine systems as well as reproductive functions. Immunotoxicity and effects on the reproductive system appear to be among the most sensitive responses.

Bibliography

Balliet RM, Chen G, Dellinger RW et al (2010) UDP-glucuronosyltransferase 1A10: activity against the tobacco-specific nitrosamine, 4-(methylnitrosamino)-1-(3-pyridyl)-1-butanol, and a potential role for a novel UGT1A10 promoter deletion polymorphism in cancer susceptibility. Drug Metab Dispos 38:484–490

Bessems JG, Vermeulen NP (2001) Paracetamol (acetaminophen)-induced toxicity: molecular and biochemical mechanisms, analogues and protective approaches. Crit Rev Toxicol 31:55–138

Birnbaum LS (1994) The mechanism of dioxin toxicity: relationship to risk assessment. Environ Health Perspect 102(suppl 9):157–167

Bu-Abbas A, Clifford MN, Ioannides C et al (1995) Stimulation of rat hepatic UDP- glucuronosyl transferase activity following treatment with green tea. Food Chem Toxicol 33:27–30

Cao Z, Li Y (2004) Potent induction of cellular antioxidants and phase 2 enzymes by resveratrol in cardiomyocytes: protection against oxidative and electrophilic injury. Eur J Pharmacol 489:39–48

Chen CH (2012) Activation and detoxification enzymes: functions and implications. Springer, New York

Conaway CC, Wang CX, Pittman B et al (2005) Phenethyl isothiocyanate and sulforaphane and their N-acetylcysteine conjugates inhibit malignant progression of lung adenomas induced by tobacco carcinogens in A/J mice. Cancer Res 65:8548–8557

Dashwood RH, Xu M, Hernaez JF et al (1999) Cancer chemopreventive mechanisms of tea against heterocyclic amine mutagens from cooked meat. Proc Soc Exp Biol Med 220:239–243

Elbekai RH, El-Kadi AO (2004) Modulation of aryl hydrocarbon receptor-regulated gene expression by arsenite, cadmium, and chromium. Toxicology 202:249–269

Furst A (2002) Can nutrition affect chemical toxicity? Int J Toxicol 21:419–424

Gooderham NJ, Murray S, Lynch AM et al (2001) Food-derived heterocyclic amine mutagens: variable metabolism and significance to humans. Drug Metab Dispos 29:529–534

Kensler TW, Groopman JD, Eaton DL et al (1992) Potent inhibition of aflatoxin-induced hepatic tumorigenesis by the monofunctional enzyme inducer 1,2-dithiole-3-thione. Carcinogenesis 13:95–100

Li X, Parkin S, Duffel MW et al (2010) An efficient approach to sulfate metabolites of polychlorinated biphenyls. Environ Int 36:843–848

Macé K, Aguilar F, Wang JS et al (1997) Aflatoxin B1-induced DNA adduct formation and p53 mutations in CYP450-expressing human liver cell lines. Carcinogenesis 18:1291–1297

Maliakal PP, Coville PF, Wanwimolruk S (2002) Decreased hepatic drug metabolising enzyme activity in rats with nitrosamine-induced tumours. Drug Metabol Drug Interact 19:13–27

Manson MM, Ball HW, Barrett MC et al (1997) Mechanism of action of dietary chemoprotective agents in rat liver: induction of phase I and II drug metabolizing enzymes and aflatoxin B1 metabolism. Carcinogenesis 18:1729–1738

Medinsky MA, Kenyon EM, Schlosser PM (1995) Benzene: a case study in parent chemical and metabolite interactions. Toxicology 105:225–233

Musavi S, Kakkar P (1998) Diazepam induced early oxidative changes at the subcellular level in rat brain. Mol Cell Biochem 178(1–2):41–46

Pegram RA, Chou MW (1989) Effect of nitro-substitution of environmental polycyclic aromatic hydrocarbons on activities of hepatic phase II enzymes in rats. Drug Chem Toxicol 12:313–326

Pfeiffer E, Diwald TT, Metzler M (2005) Patulin reduces glutathione level and enzyme activities in rat liver slices. Mol Nutr Food Res 49:329–336

Pool-Zobel B, Veeriah S, Böhmer FD (2005) Modulation of xenobiotic metabolising enzymes by anticarcinogens -- focus on glutathione S-transferases and their role as targets of dietary chemoprevention in colorectal carcinogenesis. Mutat Res 591:74–92

Shimada T (2006) Xenobiotic-metabolizing enzymes involved in activation and detoxification of carcinogenic polycyclic aromatic hydrocarbons. Drug Metab Pharmacokinet 21:257–276

Talalay P (1989) Mechanisms of induction of enzymes that protect against chemical carcinogenesis. Adv Enzyme Regul 28:237–250

Tampal N, Lehmler HJ, Espandiari P et al (2002) Glucuronidation of hydroxylated polychlorinated biphenyls (PCBs). Chem Res Toxicol 15:1259–1266

Thompson D, Oster G (1996) Terfenadine is indicated for the relief of symptoms associated with seasonal allergic rhinitis such as sneezing, rhinorrhea, pruritus, and lacrimation. JAMA 275:1339–1341

Vernhet L, Séité MP, Allain N et al (2001) Arsenic induces expression of the multidrug resistance-associated protein 2 (MRP2) gene in primary rat and human hepatocytes. J Pharmacol Exp Ther 298:234–239

Wang LQ, James MO (2006) Inhibition of sulfotransferases by xenobiotics. Curr Drug Metab 7:83–104

Wells MS, Nerland DE (1991) Hematotoxicity and concentration-dependent conjugation of phenol in mice following inhalation exposure to benzene. Toxicol Lett 56:159–166

Zhang H, Forman HJ (2009) Signaling pathways involved in phase II gene induction by alpha, beta-unsaturated aldehydes. Unsaturated aldehydes. Toxicol Ind Health 25(4–5):269–278

Chapter 3
Absorption, Metabolism, and Excretion of Foreign Compounds

Foreign compounds that humans ingest or inhale can be classified into two categories basing on their solubilities in water: hydrophiles (soluble in water) and lipophiles (soluble in lipid medium). Membrane lipid bilayers serve as physical barriers for xenobiotics transporting across cell membranes. Transport mechanism for lipophilic compounds is distinctive from hydrophilic substances. Being lipid-soluble, lipophilic molecules generally can diffuse across cell membranes, while hydrophilic substances usually are unable to penetrate lipid membranes and require carriers or channels for crossing cell membranes.

The process by which foreign compounds cross cell membranes and enter the bloodstream is referred to as absorption. Before they reach the bloodstream, xenobiotics must cross the intestinal epithelium, basement membrane and capillary endothelium. The gastrointestinal tract is one of the most important sites where foreign compounds are absorbed. When a foreign compound is administrated orally, it undergoes metabolism in the liver before reaching systemic circulation.

Lipophilic foreign compounds require metabolic conversion into hydrophilic metabolites before they are excreted from the body. Urinary and biliary systems are two primary routes for the excretion. Figure 3.1 describes various processes, including the entry, absorption, and metabolism, that foreign compounds go through before their excretion from the body.

Xenobiotics transport across the intestine includes passive diffusion and active transport. Passive diffusion, a major process for absorption of foreign compounds, is dependent on the concentration gradient and the lipid-water partition coefficient of xenobiotics, while active transport requires acquiring energy to transfer substrate (foreign compound) against a higher concentration across cell membranes.

Because lipid bilayers serve as physical barriers for biological membranes, hydrophilic foreign compounds cannot spontaneously enter into inner cell compartment. They are transported across cell membranes through mechanisms such as facilitated diffusion and active transport.

C.-H. Chen, *Xenobiotic Metabolic Enzymes: Bioactivation and Antioxidant Defense*,
https://doi.org/10.1007/978-3-030-41679-9_3

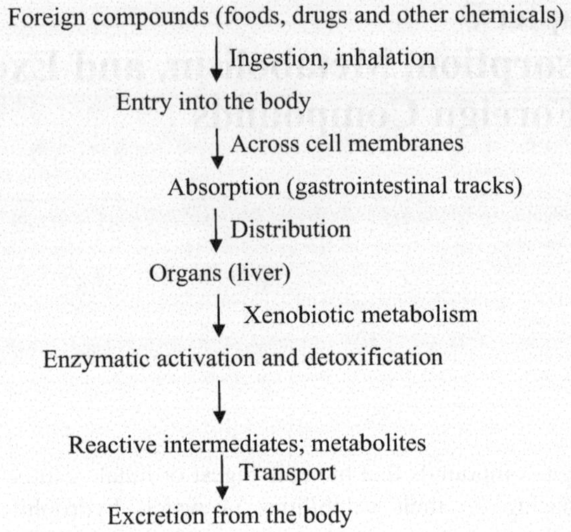

Fig. 3.1 Absorption, metabolism, and excretion of foreign compounds

3.1 Lipophiles Versus Hydrophiles

Water serves as the solvent for the intracellular and extracellular medium. Water owes its great superiority as a solvent for ionic substances partly due to its polarity: a positive hydrogen end and a negative oxygen end, giving a dielectric constant of 78. In comparison, the interior of biological membranes is a nonpolar medium with a low dielectric constant of around 4.

The polarity of water permits it to solvate ionic compounds strongly. Hydration occurs when solute compounds dissolve in water. Besides polarity, water also has the capability of forming intramolecular and intermolecular hydrogen bonding. The solubility of a foreign compound in water increases when it forms intermolecular hydrogen bonding with water.

Some foreign compounds that human ingest or inhale are hydrophilic, while some others are lipophilic. The general rule of solubility postulates that "like dissolves like." That is, nonpolar or weakly polar compounds dissolve in nonpolar or weakly polar solvents, while polar compounds dissolve in polar solvents. Thus, lipophilic foreign compounds with low dielectric constants prefer to dissolve in low dielectric constant lipid medium, but not in high dielectric constant water solvent.

Table 3.1 presents some typical lipophilic foreign compounds that humans ingest and inhale, including chemicals, drugs, and foods. The relative solubility of these foreign compounds in water is also included.

Table 3.1 Typical foreign compounds that humans ingest

Foreign compounds	Solubility in water
Chemicals	
Benzo[a]pyrene	Insoluble
Quinone	Slightly
Polychlorinated biphenyls	Insoluble
Aniline	Insoluble
Azo	Insoluble
Dimethyl phthalate	Slightly
Drugs	
Menadione	Insoluble
Diazepam	Very slightly
Acetaminophen	Very slightly
Terfenadine	Slightly
Foods	
Turmeric	Low
Caffeine	Moderate soluble
Lycopene	Insoluble

3.2 Sites of Action

The liver encounters foreign compounds such as drugs, chemicals, and foods, after they are absorbed in the intestinal tract. The liver is the major organ, where xenobiotics are metabolized, reactive intermediates are produced, and toxicity and carcinogenesis are manifested. The metabolic site for drugs depends upon the presence of foreign compound metabolizing enzymes, and most of these enzymes are present in the liver.

Hepatocytes contain activation and detoxification enzymes. Activation enzymes catalyze functionalization reactions to increase the solubility of foreign compounds in water and meanwhile generate reactive intermediates or metabolites. Detoxification enzymes catalyze conjugation reactions to detoxifying foreign compounds so as to facilitate their excretion from the body.

Although metabolic intermediates or metabolites may react at the site where they are generated, the liver is not necessarily the target organ of toxicity. The metabolites may diffuse away and react with other targets, and metabolic intermediates may be transported to other organs where they exert toxic effects. However, before transported to other organs, metabolic reactive intermediates or metabolites may potentially cause toxicity in the liver.

Besides the liver, the kidney is a frequent target organ of toxicity and is also a main site for drug metabolism. The kidney, which receives a large amount of blood, also contains a variety of foreign compound-metabolizing enzymes. The breast, lung, and colon are frequent sites of tumorigenesis in spite of their limited metabolic ability. Moreover, metabolites of xenobiotics, such as aromatic amines, are also transported to the bladder, where they are released and converted to carcinogenic species.

The accumulation of foreign compounds at the site of action is facilitated by absorption and distribution. Absorption is the transfer of a foreign compound from the site of exposure into the general circulation. Lipid solubility of foreign compounds is usually the most important property that influences their absorption. While transporters may contribute to the gastrointestinal absorption of some chemicals, a large majority of foreign compounds traverse epithelial barriers and reach blood capillaries by diffusion through cells.

3.3 Cell Membranes

Living organisms use cell membranes as hydrophobic permeability barriers to control access to the internal cell compartment. Lipid bilayers serve as physical barriers that do not favor a spontaneous exchange of foreign compounds between the internal and external cell compartments. Physical barriers of cell membranes attribute to the role of membrane permeability and hydrophobicity in the metabolism of foreign compounds.

The movement of foreign compounds into or out of cell membranes is carried out by various transport mechanisms depending on their solubility characteristics. Lipophilic, nonpolar compounds are able to move across cell membranes, while hydrophilic, polar compounds are largely restricted to the extracellular compartment and cannot enter into the cells simply by free diffusion. Consequently, the uptake of hydrophilic compounds across cell membranes is mediated by channels or transport proteins, which specifically select substrates (solutes) from the extracellular medium.

3.4 Transport Mechanisms

Major mechanisms for the transport of foreign compounds across biological membranes include passive diffusion, active transport, and facilitated diffusion. These transport mechanisms are discussed below.

3.4.1 Passive Diffusion

Passive diffusion of solutes across plasma membranes is composed of three steps: partition from the external aqueous medium to the membrane lipid phase, diffusion across membrane lipid bilayers, and partition into the internal cellular aqueous medium. In passive diffusion, the driving force for solute to move across membrane lipids into the cells is the concentration gradient, in which the concentration of solute in the external cell medium is higher than that in the internal cell medium.

3.4.2 Facilitated Diffusion

A large number of foreign compounds are transported across cell membranes through facilitated diffusion. In facilitated diffusion, the transport of a foreign compound across biomembranes into the cells is facilitated by a transport protein (carrier). Facilitated diffusion involves no input energy. It occurs downhill in accordance with the solute concentration gradient. Transport protein selects a specific solute from the extracellular medium. The binding of solute to transport protein enables the transport protein to carry the solute across cell membranes into the internal cell medium, such as in the case of glucose permeation—mediated by glucose transporter protein.

Since facilitated diffusion involves no input energy, it occurs downhill in accordance with the solute concentration gradient. Facilitating membrane permeation of inorganic ions and organic compounds also involve channels besides transporters. Channels exist in two primary states: open and closed. In the open state, channels act as pores for selected ions, allowing them to permeate across cell membranes, and then return to the closed state.

By contrast, transporter protein forms an intermediate complex with a specific solute (the substrate) on the external membrane, which induces the translocation of the substrate to the internal membrane. Transporter proteins are membrane proteins that control the influx of foreign compounds, including drugs, essential nutrients and ions, as well as the efflux of cellular wastes.

Mechanistic difference between channels and transports results in a marked difference in their turnover rates. The turnover rate constants of typical channels are much larger than those of transporter proteins.

ATP binding cassette (ABC-type transporter) and solute carrier (SLC-type transporter) are two major families of membrane transporters for drug and other foreign compounds. Membrane transporters work in concert with foreign compound-metabolizing enzymes to mediate the uptake and efflux of xenobiotics and their metabolites. Generally, SLC transporters mediate either influx or efflux of drug, while ABC transporters mediate unidirectional efflux.

3.4.3 Active Transport

Active transport of a foreign compound across bilayer membranes into the cells is also mediated by membrane transporters. But, unlike facilitated diffusion, active transport is characterized by the requirement of energy as well as the movement of solute against a concentration gradient. Depending on the driving force, active transport can be classified into primary and secondary active transports.

Primary active transport is coupled with ATP hydrolysis catalyzed by Na^+, K^+-ATPase, which provides the energy for the uptake of solute against its concentration gradient. The unidirectional movement of a solute across membranes in mammalian cells is mediated by transporter proteins such as ATP-binding cassette transporters.

In secondary active transport, the transport of solute across cell membranes uphill against its concentration gradient is coupled with the movement of another solute downhill in accordance with its concentration gradient. Therefore, secondary active transport takes place at the expense of a preexisting electrochemical gradient of another solute. In this case, the driving force for the uptake of solute across cell membranes is the electrochemical potential stored in a concentration gradient of another solute.

An example of secondary active transport is the uptake of lactose across *E. coli* membranes, which is coupled with the movement of H^+ downhill in accordance with proton electrochemical potential. Another example is Na^+- Ca^{++} exchange protein that uses the energy stored in the Na^+ gradient established by Na^+, K^+-ATPase to export cytosolic Ca^{++}.

3.5 Metabolic Pathways

Physical barriers of cell membranes attribute to the role of membrane permeability and hydrophobicity in the metabolism of foreign compounds. The capacity of excreting a wide range of lipophilic foreign compounds is the major challenge of xenobiotic metabolic pathways. This challenge is achieved by a combination of low specificity enzymatic systems in addition to physical barriers of cell membranes. Low specificity of enzymatic systems makes it possible to metabolize a wide range of lipophilic and hydrophilic compounds.

Foreign compound-metabolizing enzymes are produced based on the information stored within the genes. Xenobiotic metabolic enzymes are present in the liver at much higher concentrations than in other organs. The liver is the primary site for metabolism of foreign compounds. The liver contains the necessary enzymes for metabolism of drugs and other xenobiotics. Although liver is the primary site for foreign compound metabolism, virtually all tissue cells have some metabolic activities. Other organs that have significant metabolic activities include the gastrointestinal tract, kidneys, and lungs.

The metabolism of lipophilic foreign compounds involves a set of metabolic pathways. Two major enzyme-catalyzed pathways are (a) the functionalization in activation reactions, which involves the addition of a functional group to a xenobiotic, and (b) the conjugation in detoxification reactions, which involves the coupling of an endogenous cofactor molecule to a functionalized foreign compound.

These two pathways are responsible for the metabolism of foreign compounds, but generally they are not implicated in the elimination of endogenous metabolites derived from normal cellular constituents. Metabolic enzymes involved in activation reactions are primarily located in the endoplasmic reticulum of the liver cell, while those involved in detoxification reactions are mainly located in the cytosol.

Drugs are usually lipophilic compounds, which can pass plasma membranes and reach the site of action. Before their excretion, drugs often undergo both activation and detoxification reactions. When a drug molecule is oxidized, hydrolyzed, or covalently attached to a hydrophilic species, the molecule becomes more hydrophilic. Drug metabolism is basically a process that introduces hydrophilic functional group to the drug molecule so as to facilitate its excretion from the body.

3.5.1 *Activation Metabolism*

Activation metabolism catalyzes a functionalization reaction by means of oxidation, reduction, or hydrolysis, where a functional group is introduced to the structure of a lipophilic foreign compound (referred to as phase I metabolism). While the addition of a functional polar group results in an increase in the water solubility of the parent compound, the functionalization also activates the foreign compound, often leading to the formation of a metabolic intermediate or metabolite. Many of metabolic intermediates or metabolites are chemically reactive and are ultimately responsible for their toxicities.

3.5.2 *Detoxification Metabolism*

A functionalized foreign compound further undergoes conjugation reaction catalyzed by detoxification enzyme. The conjugation reaction combines a small cofactor molecule with the functionalized foreign compound to form a conjugate (referred to as phase II metabolism). The conjugation reaction detoxifies metabolic intermediate, rendering it less harmful, and in the meantime largely increases its water solubility, thus facilitating its excretion from the body.

Figure 3.2 illustrates two major metabolic enzymatic reactions (functionalization and conjugation) for lipophilic foreign compounds. Metabolism of hydrophilic

$$\text{(a) Lipophilic foreign compound} \xrightarrow[\text{Activation enzyme}]{\text{Functionalization}} \text{Metabolic intermediate or metabolite}$$

$$\text{(b) Metabolic intermediate + Cofactor ligand} \xrightarrow[\text{Detoxification enzyme}]{\text{Conjugation}} \text{Water soluble conjugate}$$

(c) Overall reactions:

$$\text{Lipophilic foreign compound} \xrightarrow{\text{Activation enzyme}} \text{Metabolic intermediate} \xrightarrow{\text{Detoxification enzyme}} \text{Soluble conjugate}$$

Fig. 3.2 Major metabolic pathways for lipophilic foreign compounds

foreign compounds generally can bypass functionalization reactions and directly undergo conjugation reactions.

3.6 Transport to External Cell Compartment

The next process following xenobiotic metabolisms is the transport of conjugates and other metabolites from the internal to the external cell compartment (referred to as phase III metabolism). Preceding their excretion, conjugates require transmembrane movement from the internal to the extracellular cellular compartment. Phase III metabolism of foreign compounds is the step that occurs after metabolic conversion and before their excretion from the body.

Due to membrane lipids acting as the physical barriers, the transport of conjugates or metabolites out of the cells cannot be carried out by free diffusion. Instead, the transmembrane movement of conjugates across cell membranes requires transport proteins such as export pumps. Conjugates may be further processed before being recognized by transport proteins and prior to moving out of the cells. A number of ATP-dependent transport proteins or export pumps have been identified in the liver.

ATP-binding transporters are the family of multidrug resistance proteins. In addition, vectoral transport of solute across cell membranes also plays a major role in urinary and hepatobiliary excretion of drugs from the blood to the lumen. ABC transporters are able to achieve vectoral transport by extruding lipophilic xenobiotics to the exterior compartment of cells.

3.7 Metabolism Precedes Before Excretion

Foreign compounds which are excreted from the body include (a) waste products from the digestion of foods, (b) drugs accumulated in the body, (c) chemical substances in the environment, and (d) industrial chemicals in the household. Due to limited solubility in water, lipophilic compounds require metabolic conversion into hydrophilic metabolites before they are excreted from the body. However, water-soluble, hydrophilic foreign compounds do not require metabolic conversion to increase their water solubility and can proceed directly from detoxification reaction to transport process before being excreted from the body.

Figure 3.3 illustrates metabolic processes that foreign compounds may precede before they are excreted from the body.

a. Lipophilic foreign compounds
Lipophile → Activation → Detoxification → Transport → Excretion

b. Hydrophilic foreign compounds
Hydrophile → Detoxification → Transport → Excretion

Fig. 3.3 Foreign compound metabolism precedes before excretion

3.8 Excretion

Excretion, a process in which metabolic waste products is eliminated from an organism, is primarily carried out by the kidneys, lungs and skin. Urinary and biliary systems are two primary routes for the excretion of foreign compounds and their metabolites from the body.

Renal, hepatic, and skin excretions are described below.

3.8.1 Renal Excretion

The kidney is very efficient in the elimination of toxicants from the body and is critical to the body's defense against foreign compounds. The kidney is the most important organ for the excretion of foreign compounds and their metabolites, including foods, drugs, and other chemicals. More foreign compounds are eliminated from the body by the kidney than by other organs. Renal excretion plays an important role in eliminating conjugation products resulting from functionalization reactions.

Renal excretion involves glomerular filtration, active tubular secretion, and passive tubular re-adsorption. Small compounds (<60 kD) are able to pass the glomerular filtration barrier, while larger metabolites are excreted via other pathways. After glomerular filtration, a foreign compound or its metabolite may remain in the tubular lumen and be excreted in urine. Specific carriers located on the basolateral membrane of proximal tubule cells can transport foreign compounds from the blood into the epithelial cells for subsequent excretory.

Structurally diversified organic cations and anions, including positively and negatively charged drugs and their metabolites, are secreted in the proximal tubule of the kidney. Transporters in the kidney mediate the secretion or reabsorption of many foreign compounds, thereby influencing the plasma levels of their substrates. SLC transporters and ABC transporters are two major classes of secretory transporters in the mammalian kidney.

SLC transporters are involved in moving organic cations across the basolateral membrane. They are also implicated in a variety of organic anions that are secreted in the proximal tubule. Major physiological functions of ABC transporters include the transport of toxic compounds, lipids, and bile salts. ABC transporters are also involved in the secretion of organic cations. They translocate a variety of compounds through cell membranes against the concentration gradients at the expense of ATP hydrolysis.

Besides SLC and ABC transporters, other transporters also participate in the excretion of organic ions. Transporters such as multidrug-resistance-associated proteins localized in the apical brush-border membrane are also responsible for the excretion of conjugated metabolites.

3.8.2 Reabsorption in the Kidney

Transporters in the kidney mediate the secretion or reabsorption of many foreign compounds and thereby influence the plasma levels of their substrates. Kidney is efficient in excretion of hydrophilic compounds. Hydrophilic compounds and ions are readily excreted in the urine, while lipophilic compounds that have high lipid-water partition coefficients can be re-adsorbed efficiently across the kidney tubules back into the bloodstream.

In addition to poor water solubility, tubular reabsorption in the kidney is a major factor contributing to the difficulty in the excretion of lipophilic foreign compounds. Before they are converted into water-soluble polar compounds, lipophilic compounds are not readily eliminated from the body. Therefore, living organisms develop metabolic enzyme systems that are proficient in the conversion of lipophilic into hydrophilic compounds as discussed above.

3.8.3 Hepatic Excretion

Besides being the main site for biotransformation of foreign compounds, the liver also plays an important role in removing xenobiotics from blood, after they are absorbed in the gastrointestinal tract. Foreign compounds or their conjugates with molecular weights >325 can be excreted into bile in substantial quantities. Hepatic uptake of organic anions, cations, and bile salts can be carried out by SLC transporters in the basolateral membrane of hepatocytes. SLC transporters mediate either influx or efflux of foreign compounds. This uptake process takes place by either facilitated diffusion or secondary active transport.

Moreover, ABC transporters in the canalicular membrane of hepatocytes also mediate unidirectional efflux of foreign compounds. The excretion of drugs and their metabolites mediated by ABC transporters is carried out uphill against a concentration gradient, where ATP hydrolysis provides the driving force for the transport.

3.8.4 Skin Excretion

The skin consists of an outer layer called the epidermis and an inner layer called dermis. Its primary functions are temperature control and pheromone release. Sweat glands in the skin secrete fluid waste called sweat or perspiration. Through the production of sweat by sweat glands, the skin plays a role in excretion. Sweating also maintains the level of salt in the body. However, its role as a part of the excretory system is minimal.

Bibliography

Borst P, Elferink RO (2002) Mammalian ABC transporters in health and disease. Annu Rev Biochem 71:537–592

Burckhardt BC, Burckhardt G (2003) Transport of organic anions across the basolateral membrane of proximal tubule cells. Rev Physiol Biochem Pharmacol 146:95–158

Buxton ILO, Benet LZ (2011) Pharmacokinetics: the dynamics of drug absorption, distribution, metabolism and elimination. In: Brunton LL et al (eds) Goodman & Gilman's the pharmacological basis of therapeutics. McGraw-Hill, New York

Caldwell J, Gardner I, Swales N (1995) An introduction to drug disposition: the basic principles of absorption, distribution, metabolism, and excretion. Toxicol Pathol 23(2):102–114

Chen CH (2012) Activation and detoxification enzymes: functions and implications. Springer, New York

Chen LS, Chen CH (1986) Energetic studies of lactose active transport in *Escherichia coli* membrane vesicles. Arch Biochem Biophys 246:515–524

Dean M, Rzhetsky A, Allikmets R (2001) The human ATP-binding cassette (ABC) transporter superfamily. Genome Res 11:1156–1166

Dresser MJ, Leabman MK, Giacomini KM (2001) Transporters involved in the elimination of drugs in the kidney: organic anion transporters and organic cation transporters. J Pharm Sci 90:397–421

Giacomini KM, Sugiyama Y (2011) Membrane transporters and drug response. In: Brunton LL et al (eds) Goodman and Gilman's the pharmacological basis of therapeutics. McGraw-Hill, New York

Hediger MA, Romero MF, Peng JB et al (2004) The ABCs of solute carriers: physiological, pathological and therapeutic implications of human membrane transport proteins. Pflugers Arch 447:465–468

Koepsell H (1998) Organic cation transporters in intestine, kidney, liver, and brain. Annu Rev Physiol 60:243–266

Lehman-McKeeman LD (2010) Absorption, distribution, and excretion of toxicants, Chap. 5. In: Casarett & Doull's essentials of toxicology. McGraw-Hill, New York

Chapter 4
Metabolisms of Foreign Compounds

In response to the exposure of foreign compounds such as foods, drugs, and other chemicals, humans have evolved enzyme-catalyzed metabolic mechanism to activate, detoxify, and facilitate their excretion from the body. Metabolic enzymes are involved in the biotransformation and detoxification of foreign compounds. Xenobiotic metabolism is also critically important in the elimination of metabolic intermediates and metabolites from the body. Metabolism through various enzyme systems makes the parent compounds more soluble in water for the next detoxifying step, thus facilitating their excretion from the body. As a result, xenobiotic metabolism protects and defends the body against the potential harmful foreign compounds and their reactive intermediates that are generated from enzyme-catalyzed reactions.

The conversion of lipophilic, nonpolar foreign compound to hydrophilic, polar species is essential to facilitate their excretion from the body through urine or bile. The conversion process occurs in two separate metabolic steps: (a) activation enzyme-catalyzed functionalization reactions and (b) detoxification enzyme-catalyzed conjugation reactions. In the functionalization reaction, a polar group, such as –OH or –COOH, is introduced to a foreign compound through oxidation, reduction, or hydrolysis reaction. Functionalization reaction increases the water solubility of the parent compound, which paves the way for the next metabolic step, conjugation metabolism. However, for a hydrophilic, polar foreign compound, it can bypass the functionalization reaction and directly takes part in the conjugation reaction.

In many cases, a foreign compound such as a drug may form reactive intermediate or metabolite after the introduction of a functional group. The formation of reactive intermediate or metabolite may lead to an alteration in chemical or biological reactivity of the parent compound. Such biotransformation of foreign compounds, including drugs and environmental chemicals, is commonly referred to as

---

The original version of this chapter was revised. The correction to this chapter is available at https://doi.org/10.1007/978-3-030-41679-9_20

C.-H. Chen, *Xenobiotic Metabolic Enzymes: Bioactivation and Antioxidant Defense*,
https://doi.org/10.1007/978-3-030-41679-9_4

bioactivation. Meanwhile, other potentially toxic compounds may also require biotransformation in order to exert their toxic effects.

Cytochrome P450s are the major activation enzymes involving foreign compound metabolism to form reactive intermediates or metabolites that have the capacity to interact with cellular components such as proteins, DNA, and lipids and ultimately lead to toxic effects. Such toxicants or carcinogens generated from reactive intermediates or metabolites play a major role in the pathogenesis of many disease conditions.

CYP450s are the most important enzyme family in the metabolism of the vast majority of drugs and other foreign compounds. The family of CYP450 is complex and involves a diversity of catalytic activities. Each CYP450 gene is given a number associated with a specific group within the gene group, a capital letter indicating the gene's subgroup, and a number assigned to the specific gene within the subgroup. For example, CYP450 gene that is in group 1, subgroup A, and gene 2 is written as CYP1A2. Liver microsomes contain numerous CYP450 isozymes, such as CYP1A2 and CYP2A6.

Following functionalization reactions, foreign compounds further undergo conjugation reactions catalyzed by detoxification enzymes, such as glutathione *S*-transferases, UDP glucuronosyl transferases, sulfotransferase, and *N*-acetyltransferase. A conjugate is then produced by combining a functionalized compound with an endogenous substrate (a cofactor). Conjugation can modify the reactive nature of foreign compounds to make them less toxic and more soluble in water and thus facilitate them to excrete from the body.

4.1 Activation Metabolism

As pointed out above, activation enzyme-catalyzed oxidation, reduction, or hydrolysis reaction leads to the introduction of a functional group to a foreign compound, resulting in a modification of the parent compound and an increase in its water solubility. The addition of functional groups to foreign compounds is shown in the following chemical reactions, where the symbol Φ or Φ' denotes an aromatic group, R or R' represents a straight or cyclic aliphatic group, and the equations may or may not be balanced. The product of activation metabolism subsequently serves as the substrate for detoxification enzyme-catalyzed conjugation reaction. Major reactions catalyzed by activation enzymes, including N- and O-dealkylation, aliphatic and aromatic hydroxylation, N- and S-oxidation, epoxidation, and hydrolysis, are briefly described below.

4.1.1 Functionalization Reactions

(a) N-oxidation

$$R - NH_2 \overset{O_2}{\rightarrow} R - NH - OH$$

N-Oxidation results in forming hydroxylamine, such as the conversion of primary and secondary amines to give hydroxylamines. Other examples include N-oxidation of aromatic amines such as procainamide, an anti-arrhythmic drug, and imides such as phenytoin, an anti-epileptic drug.

(b) S-oxidation

$$\begin{array}{ccc} R' & & R' \\ | & O_2 & | \\ R\text{-}SH_2 & \rightarrow & R\text{-}S\text{=}O \end{array}$$

S-Oxidation results in forming sulfoxide or sulfone, such as S-oxidation of chlorpromazine, an anti-psychotic medication.

(c) Hydroxylation

$$R - \Phi \overset{OH}{\rightarrow} R - \Phi - OH$$
$$R - \Phi - CH_3 \overset{OH}{\rightarrow} R - \Phi - CH_2 - OH$$

Aromatic or aliphatic hydroxylation results in generating phenyl or hydroxyl compound, such as the conversion of amino acid proline residue to hydroxyproline.

(d) O-dealkylation

$$R - O - CH_3 \rightarrow ROH + HCHO$$

O-Dealkylation results in the loss of alkyl group attached to oxygen, such as the conversion of encainide (an antiarrhythmic agent) to *O*-demethylencainide.

(e) N-dealkylation

$$\Phi - N - CH_3 \rightarrow \Phi - NH_2 + CH_2OH$$
$$\Phi - O - CH_3 \rightarrow \Phi - OH + CH_2OH$$

N-Dealkylation is a common metabolic reaction for drugs containing secondary and tertiary amines. N-Dealkylation results in the loss of alkyl group attached to nitrogen or oxygen in the compound, such as N-dealkylation of 9-alkylaminoacridine (an anti-malarial drug).

(f) Epoxidation

$$\begin{array}{ccc} \text{-}\Phi\text{-RC=CR'} & \rightarrow & \text{-}\Phi\text{-RC-CHR'} \\ & & \diagdown_{O}\diagup \end{array}$$

Epoxidation is a pathway in metabolizing many carcinogenic aromatic hydrocarbons, such as the conversion of aflatoxin to aflatoxin-epoxide. Aflatoxin is a toxin produced by fungi that are found in agricultural crops such as corn.

(g) Nitro-reduction

$$-\Phi - NO_2 \rightarrow -\Phi - NH_2$$

Reduction of nitro compounds to form amide, such as the conversion of nitrobenzene to benzamine (aniline).

(h) Amide-reduction

$$RCO - NR'R'' \rightarrow RCH_2NR'R''$$

Reduction of amide to form amine compound, such as the conversion of sulfanilamide (an antibacterial) to its hydroxylamine.

(i) Hydrolysis

$$\Phi - CO - NH_2 + H_3O^+ \rightarrow \Phi - COOH + NH_4{}^+$$
$$R - COOR' + H_2O \rightarrow RCOOH + R' - OH$$

Hydrolysis is a pathway for a number of esters or amides, such as the conversion of propanil (a herbicide) to dichloroaniline.

Activation enzyme-catalyzed reaction leads to the introduction of a functional group such as –OH, –COOH, –SH, –O, or $-NH_2$ to a lipophilic foreign compound, resulting in increasing the water solubility of the parent compound. The modified compound serves as the substrate for conjugation reaction catalyzed by detoxification enzymes. Table 4.1 summarizes typical functional groups that are introduced through activation enzyme-catalyzed functionalization reactions.

4.1.2 Metabolic Reactive Intermediates

In many cases, lipophilic compounds are metabolized and converted to form activated intermediates during activation metabolism. A chemically active intermediate is a potentially more reactive and toxic species than the parent compound. Generally, reactive intermediates are generated through functionalization reactions mediated by

Table 4.1 Typical functional groups introduced by activation enzyme-catalyzed reactions

Functionalization reaction	Functional group introduced
N-oxidation	N–OH
S-oxidation	S=O
Hydroxylation	Φ–OH
Nitro-reduction	NH_2
Amide reduction	NRR′
O-dealkylation	R–OH
N-dealkylation	NH_2
Hydrolysis	COOH
Epoxidation	C - C \ / O

CYP450s and other phase I activation enzymes. CYP450 isoenzymes are most implicated in activation enzyme-mediated bioactivation. In minor cases, conjugation reactions catalyzed by phase II detoxification enzymes are also involved.

Chemically active intermediates are capable of interacting with cellular components (proteins, lipids, and DNA) and are ultimately responsible for the toxicity of foreign compounds. A typical example is the transformation of 4-ipomeanol (a pulmonary pre-toxin) catalyzed by CYP4B1 to form an active intermediate, α,-β-unsaturated dialdehyde, which exhibits toxicity and leads to 4-ipomeanol-induced lung injuries.

Highly active intermediates of foreign compounds generated by activation metabolism also include epoxides, carbonium ions, and hydrogen peroxide. Furthermore, chemically active intermediates can induce toxicity or carcinogenicity by interacting with molecular oxygen to form reactive oxygen species such as hydroxyl (OH•) and superoxyl anion (O_2•) radicals. Reactive oxygen species, especially free radicals, are believed to be involved in degenerative diseases and cancers.

4.2 Detoxification Metabolism

Conjugation reactions lead to inactivate potentially toxic reactive intermediates that are generated in the above functionalization reactions and often also improve water solubility of foreign compounds, thus facilitating the elimination of xenobiotics from the body. A conjugate is formed by the combination of the substrate from the product of activation functionalization reaction with a small endogenous molecule (a cofactor). Conjugation reactions are mainly catalyzed by transferase enzymes.

Metabolic pathways in detoxification metabolism include glucuronide, glutathione, sulfate, and amino acid conjugation reactions. Conjugation reactions usually require the substrate (foreign compound) to have an oxygen, nitrogen, or sulfur atom which serve as the acceptor site for the hydrophilic moiety, such as glutathione, glucuronic acid, sulfonate, or acetyl group.

4.2.1 *Conjugation Reactions*

Major conjugation reactions including glucuronide, sulfate, glutathione, acetyl, methyl, and amino conjugations are briefly described below, where symbol Φ or Φ′ denotes an aromatic side group, which may be replaced with an aliphatic side group (R or R′). The equations in the following reactions may or may not be balanced.

(a) Glucuronide conjugation

$$-\Phi - \mathrm{OH} \overset{\text{UDP-glucuronic acid}}{\longrightarrow} -\Phi - \mathrm{O} - \text{glucuronide}$$

Glucuronide conjugation reaction transfers a glucuronic group from UDP-glucuronic acid to a functionalized foreign compound to form a glucuronide metabolite, such as the conversion of hydroxydiazepam (sleep pill) to hydroxydiazepam glucuronide. The functional group for glucuronide conjugation includes phenol and aromatic amines.

(b) Glutathione conjugation

$$HO - \Phi - OH + GSH \rightarrow HO - \Phi - GSH$$

Glutathione (GSH) is a tripeptide (glutamate-cysteine-glycine) containing a sulfhydryl group. Glutathione conjugation transfers glutathione to a functionalized foreign compound, such as the conversion of paracetamol (pain reliever) to paracetamol glutathione.

(c) Sulfonate conjugation

$$-\Phi - OH + PAPS \rightarrow -\Phi - O - SO_2 - OH + PAP$$

Sulfonate conjugation is an important pathway in the metabolism of an amino group of foreign compounds. Sulfonate conjugation transfers sulfonate from 3′-phosphoadenosine-5′-phosphosulfate (PAPS) to the –OH group of aromatic or aliphatic compound, such as the conversion of hydroxycoumarin (anticoagulant) to hydroxycoumarin sulfate.

(d) Amino acid conjugation

$$-\Phi - COOH + H_2NCH_2COOH \rightarrow -\Phi - CONHCH_2COOH + H_2O$$

Amino acid conjugation reaction is important in the metabolism of an organic acid. The carboxylic group of an organic acid conjugates with an amino acid, such as the conversion of benzoic acid to hippuric acid (benzoyl amidoacetic acid). Glycine is the most common amino acid in amino acid conjugation reaction.

(e) N-acetyl conjugation

$$-\Phi - NH_2 + \text{acetyl-CoA} \rightarrow -\Phi - \text{NH-acetyl} + \text{CoA}$$

N-Acetyl conjugation is important in the metabolism of drugs and environmental chemicals containing an aromatic amine or hydrazine group. N-Acetyl conjugation transfers acetyl from cofactor acetyl-CoA to $-NH_2$ group of the metabolite, such as the conversion of $-NH_2$ on dapsone (skin disorder medication) to –NH-acetyl.

(f) Methyl conjugation

$$-\Phi - R - NH_2 \overset{SAM}{\rightarrow} -\Phi - R - NH - CH_3$$

S-Adenosyl methionine (SAM) is a common substrate involved in methyl group transfer. Methyl conjugation transfers methyl group from SAM to $-NH_2$ group

Table 4.2 Cofactors and conjugation groups in conjugation reactions

Conjugation reaction	Cofactor	Conjugation group
Glucuronidation	UDP-glucuronic acid	Glucuronic acid
Glutathione conjugation	Glutathione	Glutathione
Sulfonation	3′-Phosphoadenosine-5′-phosphosulfate	Sulfonate
Amino acid conjugation	Glycine	Amino acid
Acetylation	Acetyl-coenzyme A	Acetyl group
Methylation	*S*-Adenosyl-methionine	Methyl group

of the metabolite such as the conversion of $-NH_2$ group on dopamine (a neurotransmitter) to $-NH-CH_3$.

Table 4.2 presents typical cofactors and conjugation groups in detoxification metabolism of functionalized foreign compounds, where detoxification enzyme-catalyzed reaction introduces a conjugation group, such as glutathione, glucuronic acid, sulfonate, methyl, and acetyl, to a functionalized foreign compound, thus detoxifying foreign compound, increasing its water solubility, and facilitating its excretion from the body.

4.2.2 *Non-conjugation Reactions*

Besides the above discussed conjugation reactions, in some classification, detoxification metabolism also includes non-conjugation reactions catalyzed by detoxification enzymes such as quinone reductase and epoxide hydrolase. These non-conjugation reactions are presented below:

(a) Quinone reductase

$$R - \Phi = 0 \overset{\text{Quinone reductase}}{\longrightarrow} R - \Phi - OH$$

Quinones are reduced to hydroquinones catalyzed by NADPH-quinone oxidoreductase, such as the conversion of menadione (a nutritional supplement in animal feed) to hydroquinone, where the produced hydroquinone can further undergo glucuronidation or sulfonation.

(b) Epoxide hydrolase

$$\begin{array}{c} \Phi\text{–HC–CH}_2 \\ \backslash\ / \\ \text{O} \end{array} \overset{\text{Epoxide hydrolase}}{\longrightarrow} \begin{array}{c} \Phi\text{–HC–CH}_2 \\ |\quad\ | \\ \text{HO}\ \ \text{OH} \end{array}$$

Many epoxides are metabolic intermediates formed during CYP450-catalyzed oxidation of unsaturated aromatic or aliphatic compounds. Epoxide hydrolase catalyzes the reduction of epoxide to glycol, such as the conversion of styrene 7,8-epoxide to styrene 7,8-glycol. Epoxide hydrolase plays an important role in detoxifying electrophilic epoxides.

4.3 Defenses Against Metabolites Generated from Bioactivation

Antioxidants and detoxifying enzymes are the body's defensive mechanisms against foreign compounds or their metabolites generated from bioactivation. Antioxidant defensive mechanism includes enzymatic antioxidants and nonenzymatic antioxidants. Detoxifying enzymes catalyze the conjugation reactions that combine functionalized foreign compounds with enzyme cofactors to form conjugates, thus inactivating potentially toxic reactive intermediates or metabolites and facilitating their excretion from the body.

4.3.1 Antioxidants

(a) Enzymatic antioxidants

Major enzymatic antioxidants include superoxide dismutase, catalase, and glutathione peroxidase. These antioxidants commonly require NADPH as a reducing equivalent. Superoxide is the primary reactive oxygen species produced from the metabolic conversion of foreign compounds. Its dismutation by superoxide dismutase is of primary importance to cell defense. Superoxide dismutase enzymes are widely expressed in the lung.

Hydrogen peroxide produced by the catalytic action of superoxide dismutase is reduced to form water by catalase and glutathione peroxidase. Catalase exists as a tetramer, and each monomer contains a heme group at the active site. Glutathione peroxidase belongs to a family of tetrameric enzymes that use low molecular weight thiol such as GSH to reduce hydrogen peroxide to the corresponding alcohol.

(b) Nonenzymatic antioxidants

Nonenzymatic reactions utilize endogenous compounds such as glutathione to prevent damages to cellular components caused by peroxides and free radicals. Glutathione is abundant in cell compartments and is the major soluble antioxidant, which detoxifies hydrogen peroxide and reduces it to water and oxygen. Moreover, glutathione is a cofactor for detoxifying enzyme glutathione transferase.

Small antioxidant molecules are such as vitamins C and E. Vitamin C provides intracellular and extracellular aqueous-phase antioxidant capacity primarily by scavenging oxygen free radicals. Vitamin E is concentrated in the hydrophobic interior of cell membranes and is the main defense against oxidant-induced membrane injury.

4.3.2 Metabolic Detoxifying Enzymes

Metabolic conversion of reactive intermediates to metabolites that are no longer capable of causing cellular damages is referred to as deactivation. Conjugation reactions generally yield stable, nonreactive metabolites that are ready for excretion from the body. In minor cases, conjugation reaction may be implicated in the toxicity of foreign compounds, such as glutathione conjugation of epoxide that exhibits toxic effects on the kidney.

The amount of reactive intermediate or metabolite present in the body largely depends on the competing pathways between activation mechanism and detoxification mechanism. The human body requires a delicate balance between the rates of activation and detoxification. Any factors that interrupt this delicate balance will affect the fate of foreign compounds and their associated toxicity. When the activation pathway overwhelms the detoxification pathway, an overproduction of reactive intermediates or metabolites occurs, leading to their accumulations in the body. As a result, a potential increase in their interactions with cellular components (proteins, lipids, and DNA) occurs.

Furthermore, a deficiency in the detoxification pathway may lead to an increased vulnerability to the toxicity of foreign compounds. A low expression of detoxifying enzyme activity can result in a low rate of conjugation reaction. When the rate of utilization exceeds the rate of cofactor supply, an insufficiency of cofactors, such as glutathione, sulfate, or glucuronic acid, can also lead to a deficiency in the detoxification pathway.

Environmental factors can also attribute to the activation pathway, making it more dominant over the detoxification pathway. Large exposures to smoke, alcohol, or drug can selectively induce activation enzymes and suppress detoxification enzymes, leading to an imbalance between the rates of activation and detoxification pathways.

Bibliography

Cavalieri EL, Rogan EG (2005) The approach to understanding aromatic hydrocarbon carcinogenesis. The central role of radical cations in metabolic activation. Pharmacol Ther 55:183–199

Chen CH (2012) Activation and detoxification enzymes: functions and Implications. Springer, New York

Cribb AE, Spielberg SP, Griffin GP (1995) N4-hydroxylation of sulfamethoxazole by cytochrome P450 of the cytochrome P4502C subfamily and reduction of sulfamethoxazole hydroxylamine in human and rat hepatic microsomes. Drug Metab Dispos 23(3):406–414

Cribb AE, Peyrou M, Muruganandan S et al (2005) The endoplasmic reticulum in xenobiotic toxicity. Drug Metab Rev 37:405–442

Decker M, Arand M, Cronin A (2009) Mammalian epoxide hydrolases in xenobiotic metabolism and signalling. Arch Toxicol 83(4):297–318
Gibson GG, Skett P (2001) Introduction to drug metabolism, 3rd edn. Nelson Thornes Publishers, Cheltenham
Hayes JD, Flanagan JU, Jowsey IR (2005) Glutathione transferases. Annu Rev Pharmacol Toxicol 45:51–88
Horton AA, Fairhurst S (1987) Lipid peroxidation and mechanisms of toxicity. Crit Rev Toxicol 18 (1):27–79
Ioannides C (2002) Xenobiotic metabolism: an overview. In: Ioannides C (ed) Enzyme systems that metabolize drugs and other xenobiotics. Wiley, New York
Jana S, Mandlekar S (2009) Role of phase II drug metabolizing enzymes in cancer chemoprevention. Curr Drug Metab 10(6):595–616
Meisel P (2002) Arylamine N-acetyltransferases and drug response. Pharmacogenomics 3:349–366
Negishi M, Pedersen LG, Petrotchenko E et al (2001) Structure and function of sulfotransferases. Arch Biochem Biophys 390:149–157
Oesch F (1980) Influence of foreign compounds on formation and disposition of reactive metabolites. Ciba Found Symp 76:169–189
Park BK, Kitteringham NR, Maggs JL et al (2005) The role of metabolic activation in drug-induced hepatotoxicity. Annu Rev Pharmacol Toxicol 45:177–202
Pumford NR, Halmes NC (1997) Protein targets of xenobiotic reactive intermediates. Annu Rev Pharmacol Toxicol 37:91–117
Ram N, Kalász H, Adeghate E, Darvas F, Hashemi F, Tekes K (2012) Medicinal chemistry of drugs with active metabolites (N-, O-, and S-desalkylation and some specific oxidative alterations). Curr Med Chem 19(33):5683–5704
Rushmore TH, Kong AN (2002) Pharmacogenomics, regulation and signaling pathways of phase I and II drug metabolizing enzymes. Curr Drug Metab. 3(5):481–490
Talalay P (1989) Mechanisms of induction of enzymes that protect against chemical carcinogenesis. Adv Enzyme Regul 28:237–250
Tukey RH, Strassburg CP (2000) Human UDP-glucuronosyltransferases: metabolism, expression, and disease. Annu Rev Pharmacol Toxicol 40:581–616
Weinshilboum RM, Otterness DM, Szumlanski CL (1999) Methylation pharmacogenetics: catechol O- methyltransferase, thiopurine methyltransferase, and histamine N-methyltransferase. Annu Rev Pharmacol Toxicol 39:19–52

Chapter 5
Bioactivation Metabolism: Activation Enzymes

Biochemical reactions in living cells exhibit kinetic barriers that prevent the reactions from occurring spontaneously. A variety of enzymes that act as catalysts make the reactions energetically favorable. Consequently, biochemical reactions can occur at the rates required for maintaining cell functions. Enzymes are highly effective catalysts for an enormous diversity of biochemical reactions. They make up the largest and most highly specialized class of protein molecules.

Depending on the nature of biochemical reactions, enzymes can be divided into six major classes: (1) oxidoreductases that catalyze the oxidation and reduction reactions, (2) transferases that catalyze the transfer of functional groups, (3) hydrolases that catalyze the hydrolysis reactions, (4) lyases that catalyze the reactions involving the removal or addition of a group to double bonds, (5) isomerases that catalyze the reactions involving the intramolecular rearrangement, and (6) ligases that catalyze the reactions that join together two molecules. Foreign compound-metabolizing enzymes (activation enzymes and detoxification enzymes) belong to three of the above six classes: oxidoreductases, transferases, and hydrolases.

The substrate which the enzyme exerts its catalytic action is a small chemical molecule of foreign compound. Foreign compound-metabolizing enzymes are primarily present in the endoplasmic reticulum of the cells and/or in the cytosol. They are produced from the information stored within the genes and are present in most tissues with the highest levels located in the liver and intestines.

Activation enzymes are composed of oxidases, hydrolases, and reductases. Oxidases include CYP450s, flavin-containing monooxygenases, amine oxidases, lipoxygenases, aldehyde and xanthine oxidases, alcohol dehydrogenases, and peroxidases. Reductases include nitroreductases and azoreductases. Hydrolases include carboxylesterases and epoxide hydrolases. In the metabolism, activation enzymes that catalyze functionalization reactions to increase the water solubility of foreign compounds are referred to as activators.

C.-H. Chen, *Xenobiotic Metabolic Enzymes: Bioactivation and Antioxidant Defense*,
https://doi.org/10.1007/978-3-030-41679-9_5

5.1 Major Activation Enzymes

A wide range of foreign compounds are lipophilic compounds that are insoluble in water but are capable of passing through cell membranes and entering into cells. An ultimate goal of foreign compound metabolic enzymes is to convert lipophilic foreign compounds into water-soluble species, so as to facilitate their elimination from the body.

Three major groups of activation enzymes, oxidases, reductases, and hydrolases, catalyze oxidation, reduction, and hydrolysis reactions, respectively. Among them, there are more oxidative enzymes than reductive or hydrolytic enzymes. Enzymatic reduction reactions are relatively less investigated as compared to oxidation and hydrolysis reactions. Hydrolases catalyze hydrolysis reactions by the introduction of water.

Activation enzymes catalyze functionalization reactions by the introduction of functional groups to foreign compounds through oxidation, reduction, or hydrolysis reactions catalyzed by oxidoreductases or hydrolases, respectively. In many cases, the addition of a functional group results in the formation of an intermediate which is more chemically active and more toxic than the parent compound. Such chemical reactive intermediates or metabolites are able to react with cellular molecules (proteins, lipids, and DNA), causing toxicity to cells and organs.

Activation enzyme-catalyzed reactions may act on different atoms or groups of foreign compounds, including carbon, nitrogen, oxygen, sulfur, unsaturated hydrocarbon, ester, amide, and epoxide. Catalytic actions of activation enzymes on specific atoms or groups are presented in this chapter, while the detailed reactions catalyzed by individual activation enzymes are discussed in the following chapter. Table 5.1 lists various activation enzymes and their catalyzed reactions.

5.2 Oxidative Enzymes

Oxidative enzymes include cytochrome P450, flavin monooxygenase, amine oxidases, lipoxygenases, alcohol dehydrogenase, aldehyde oxidase, xanthine oxidases, peroxidase, and prostaglandin H synthase as presented in Table 5.1. Among the listed oxidative enzymes, alcohol dehydrogenase, aldehyde oxidase, and CYP450 can catalyze both oxidative and reductive reactions, depending on the substrate or the condition. A large majority of organic chemicals undergo oxidation reactions during metabolism.

Foreign compounds containing an aromatic ring (such as biphenyl) can undergo aromatic hydroxylation reaction, leading to the formation of phenolic products. Other foreign compounds that contain nitrogen may be oxidized to form nitrooxides or hydroxylamines. Many drugs are metabolized through aliphatic or aromatic hydroxylation. The functional properties of major oxidative enzymes are briefly discussed below.

Table 5.1 Activation enzymes and their catalyzed reactions

Activation enzymes	Catalyzed reactions
(a) Oxidative enzymes	
Cytochrome P450 oxidase	Hydroxylation, dealkylation, deamination, epoxidation, ester cleavage
Flavin-monooxygenase	N-, S-, and P-compound oxidation, N-hydroxylation
Peroxidase	Peroxidation, dealkylation, demethylation
Amine oxidase	Oxidation of amines
Lipoxygenase	Oxidation of fatty acids, N-dealkylation, epoxidation
Alcohol dehydrogenase	Oxidation of alcohols
Aldehyde oxidase	Oxidation of aldehydes
Xanthine oxidase	Oxidation of aldehydes
Prostaglandin H synthase	Peroxidation, oxidation of phenols and amines
(b) Reductive enzymes	
Nitroreductase	Reduction of nitro compounds
Azoreductase	Reduction of azo compounds
Aldo-keto reductase	Reduction of aldehydes and ketones
Carbonyl reductase	Reduction of carbonyls and quinones
(c) Hydrolytic enzymes	
Carboxylesterase	Hydrolysis of esters and amides
Epoxide hydrolase	Hydrolysis of epoxide

5.2.1 *Cytochrome P450 Oxidase*

Among foreign compound-metabolizing enzymes, cytochrome P450 (CYP450) oxidase is the most important and extensively studied activation enzyme. CYP450 refers to a unique family of heme proteins. P450 is designated because its reduced form binds carbon monoxide to produce a maximum optical absorption around 450 nm. CYP450 enzymes are hemeproteins. These enzymes contain a heme that is bound to the polypeptide chain. The family of CYP450s is complex and diverse in catalytic activities. There are close to 60 of CYP450 isozymes that have been identified in humans and are present predominantly in the liver but also occur in other organs such as the intestine, lung, and kidney.

CYP450 oxidase functions as a monooxygenase which catalyzes the insertion of one atom of oxygen molecule into the substrate. The enzymes utilize O_2 and H^+ (from NADPH) to carry out the oxidation of a foreign compound. A molecule of oxygen initially binds to the heme moiety and subsequently is cleaved. One oxygen atom forms by chemical binding with the substrate. The remaining oxygen atom is reduced to yield water. Two electrons that stem from NADPH also participate in CYP450-catalyzed reactions, where the reaction products are oxidized substrate and water.

CYP450s catalyze a diversity of foreign compounds including drugs, food additives and industrial and environmental chemicals. For instances, CYP1A1 is particularly active towards polycyclic aromatic hydrocarbons, while CYP1A2 activates aromatic amines and amides and CYP2E1 activates aflatoxin B1 and N-nitrosamines. Besides increasing the solubility of foreign compounds, the reactions may also involve in the activation of potentially toxic foreign compounds to reactive intermediates as well as the formation of many electrophilic metabolites.

The family of CYP450s catalyzes the metabolism of a large number of structurally diverse chemicals, including N- and O-dealkylation, aliphatic and aromatic hydroxylation, N- and S-oxidation, deamination, cleavage of esters, and epoxidation of a double bond. A diversity of foreign compounds catalyzed by CYP450s include foods, drugs, and industrial and environmental chemicals. In addition, CYP450s also catalyze the reactions involving endogenous compounds, such as cholesterol, steroid hormones, and fatty acids.

Microsomal CYP450 isozymes involved in metabolizing foreign compounds are responsible for the breakdown of medications, which occurs mostly in the liver. CYP450 isozymes are very important for the pharmaceutical industry and have far-reaching implications in medicine, especially in the activation of therapeutic agents. CYP450s account for about 70–80% of enzymes involved in drug metabolism.

Moreover, CYP450s play a critical role on adverse effects of foreign compounds during activation metabolism. In many cases, CYP450s catalyze the conversion of foreign compounds to reactive intermediates and metabolites, which are more chemically active and more toxic than the parent compounds and are able to react with cellular molecules (proteins, lipids, and DNA), causing toxicity to cells and organs.

Genetic factors play a role in the expression of CYP450 activities. There are significant variations in the levels of enzyme expression present in CYP450 isozymes among individuals, due to genetic polymorphisms. Genetic variants in CYP450 have a potentially functional impact on the efficacy and adverse effects of drugs. CYP450 expression in its catalytic reactions can also differ markedly as a result of exposure to environment and lifestyle.

5.2.2 *Flavin-Containing Monooxygenase*

Flavin-containing monooxygenases (FMO) are another superfamily of activation enzyme involved in the metabolism of foreign compounds. FMOs are oxidative enzymes that catalyze oxidative metabolism of nitrogen-, sulfur-, and phosphorus-containing foreign compounds, such as amines, sulfides, and phosphates. FMOs are expressed at a high level in the liver and are bound to the endoplasmic reticulum.

The reactions catalyzed by FMOs involve one-step two-electron substrate oxygenation, as opposed to two sequential one-electron oxidations that are catalyzed by CYP450. FMO-catalyzed reactions usually convert foreign compounds into relatively polar metabolites, thus facilitating their excretion from living organisms. While FMO-catalyzed oxidation reactions generally convert foreign compounds into polar, harmless metabolites, however, in some cases, FMOs activate xenobiotics into reactive intermediates that can have potential toxic effects.

The family of FMOs (FMO1, 2, 3, and 4) consists of a group of enzymes that catalyze reactions through the binding of cofactor flavin. Kinetics studies revealed that the mechanism of FMO isozymes are similar but differ in the substrate specificities. Although many advances have been made in the understanding of CYP450-catalyzed activation of drugs and other chemicals, limited information is available for FMO-mediated activation metabolism. Since amines and sulfides are prevalent in drugs, FMO-catalyzed reactions could play an important role in pharmaceutical industries.

5.2.3 *Amine Oxidase*

Amine oxidases are enzymes that catalyze the oxidation of a wide range of xenobiotic amines into aldehydes. These enzymes are widely distributed throughout the body including the liver, intestines, and other organs. Amine oxidases are involved in the oxidative deamination of primary, secondary, and tertiary amines. Among amino oxidases, monoamine oxidases (MAOs) are most extensively studied in terms of their involvement in foreign compound metabolism.

Hepatic monoamine oxidases play a key defense role in the metabolism of amines. MAOs are classified as flavoproteins and are composed of two structurally related flavin-containing enzymes (MAO-A and MAO-B). MAO-A and MAO-B are mitochondrial outer membrane-bound flavoproteins that catalyze the oxidative deamination of amines. Oxygen is used to remove an amine group from a molecule in MAO-catalyzed oxidation of aliphatic and aromatic amines. The oxygen that is incorporated into amine is derived from water.

Foreign compounds which are metabolized by monoamine oxidases include tyramine in foods and beverages, 2-phenylethylamine in dietary sources, benzylamine in mouth washes, and 2-phenylpropanolamine in decongestants and cough medicines. The products of these MAO-catalyzed reactions are ammonia, hydrogen peroxide, and aldehyde, which are potentially toxic. The produced aldehyde may be further metabolized by aldehyde dehydrogenase or aldehyde oxidase to form the corresponding carboxylic acid.

5.2.4 Lipoxygenase

Lipoxygenases play a role as major activation enzymes in the oxidation of foreign compounds. The enzymes are present in many mammalian tissues including liver, lung, kidney, and colon. Lipoxygenases belong to a family of iron-containing, non-heme enzymes, which catalyze the insertion of oxygen into poly-unsaturated fatty acids in lipids, such as arachidonic acid and linoleic acid. Unlike CYP450s that involve inserting one oxygen atom into the substrate, lipoxygenases catalyze oxidation of arachidonic acid by inserting two oxygen atoms, in which the redox state of active site iron changes from Fe^{3+} to Fe^{2+}.

Lipoxygenases catalyze oxidative metabolism of foreign compounds including drugs and other chemicals. Lipoxygenase-catalyzed reactions involve a fatty acid substrate with two double bonds separated by a methylene group. Besides catalyzing the oxygenation of polyenoic fatty acid to corresponding lipid hydroperoxides, lipoxygenases also catalyze N-dealkylation and epoxidation. Typical lipoxygenase-catalyzed N-demethylation of drugs includes aminopyrine and chlorpromazine. Lipoxygenases are also important in catalyzing the epoxidation of polycyclic aromatic hydrocarbons such as benzo[a]pyrene.

5.2.5 Alcohol Dehydrogenase

Alcohol dehydrogenases (ADHs) are oxidative enzymes composed of a group of several isozymes. These enzymes break down alcohols and are probably the most important dehydrogenases involved in the metabolism of alcohols. ADHs catalyze the oxidation of alcohols, using NAD^+ or $NADP^+$ as the electron acceptor. The reaction results in the oxidation of primary and secondary alcohols to the corresponding aldehyde or ketone by the transfer of a hydride anion to NAD^+ with the release of a proton (reduction of NAD^+ to NADH).

ADH-catalyzed oxidation reactions serve to break down alcohols and participate in generation of aldehyde and ketone as metabolites. The conversion product, aldehyde, is usually toxic. Further oxidation of aldehyde to acid is a vital detoxification reaction. ADH and aldehyde dehydrogenase are the primary enzymes involved in metabolism of alcohol. Both enzymes occur in several forms that are encoded by different genes with different characteristics that affect the level of alcohol consumption and risk of alcoholism.

5.2.6 Aldehyde Dehydrogenase

The primary enzymes involved in alcohol metabolism are alcohol dehydrogenase (ADH) and aldehyde dehydrogenase (ALDH). The major metabolic consequence of ethanol metabolism catalyzed by ADH is the production of acetaldehyde, a highly reactive intermediate. Aldehyde dehydrogenases are a group of enzymes that catalyze the oxidation of aldehydes and convert aldehydes to carboxylic acids, where the oxygen comes from a water molecule.

5.2.7 Aldehyde Oxidase

Aldehyde oxidase, a molybdozyme, is important in drug oxidation, activation, and detoxification. The enzyme exhibits oxidase activity toward various aliphatic and aromatic aldehydes. High levels of aldehyde oxidase activity are present in the liver. Unlike CYP450, the ultimate source of oxygen inserted into a substrate is water, not O_2. While aldehyde oxidase catalyzes the oxidation of aldehyde foreign compounds, the enzyme is also involved in the reduction reactions, such as the reduction of nitroaromatics to hydroxylamines.

5.2.8 Xanthine Oxidase

Xanthine oxidase is an enzyme that catalyzes the oxidation of hypoxanthine to *xanthine* and of xanthine to uric acid. Similar to aldehyde oxidase, xanthine oxidase is a molybdozyme but differs in substrate specificities. Besides the liver, a high level of xanthine oxidase activity is found in other tissues. Xanthine oxidase exhibits oxidase activity toward various aliphatic and aromatic aldehydes. Many features of aldehyde oxidase apply to xanthine oxidase. Xanthine oxidase is also important in oxidation of drugs.

5.2.9 Peroxidase

Peroxidases are a hemoprotein that catalyzes the oxidation of a number of substrates such as the breakdown of hydrogen peroxide. Hydrogen peroxide is one of reactive intermediates produced as a by-product of oxidative reactions. Peroxidase enzymes such as chloroperoxidase and horseradish peroxidase catalyze peroxide-dependent reactions in the metabolism of a wide variety of foreign compounds through the reactions that require NADPH and O_2. Horseradish peroxidase also catalyzes the reaction involving hydroperoxide-dependent N-demethylation of secondary and tertiary arylamines.

5.2.10 Prostaglandin H Synthase

Prostaglandin H synthase is a dual-function enzyme consisting of a peroxidase and a cyclooxygenase. The enzyme has two distinct catalytic activities: a peroxidase activity and a cyclooxygenase activity. Prostaglandin H synthase is one of the most extensively studied enzymes involving metabolic activation of foreign compounds, in particular in tissues with low CYP450 activity. Its peroxidase activity catalyzes the oxidation of drug such as acetaminophen. In addition to peroxide oxidation, prostaglandin H synthase also catalyzes a number of diverse oxidation reactions involving phenols and aromatic amines.

5.3 Oxidation at Specific Atom or Group

The oxidative enzyme-catalyzed reactions presented below include (1) the oxidation at carbon atom, (2) the oxidation at nitrogen atom, and (3) the oxidation of unsaturated hydrocarbon.

5.3.1 Oxidation at Carbon Atom

Oxidation reactions at carbon atom can occur on aromatic ring, alcohol, aldehyde, or another functional group of a foreign compound. Activation enzymes involved in these oxidation reactions include CYP450s, alcohol dehydrogenases, aldehyde dehydrogenases, xanthine oxidases, and peroxidases. In CYP450-catalyzed reaction, an oxygen atom is inserted in a C–H bond of an aromatic compound to form phenol.

Alcohol dehydrogenase-catalyzed oxidation reaction converts aliphatic, aromatic, or cyclic alcohol to aldehyde by removing two hydrogens (one attached to carbon and another attached to oxygen). Aldehyde can be further oxidized to form carboxylic acid. In aldehyde oxidase-catalyzed reaction, an oxygen atom is inserted in C–H bond of aldehyde to yield carboxylic acid. In peroxidase-catalyzed oxidation reaction, alcohol is converted to peroxide.

5.3.2 Oxidation at Nitrogen Atom

Oxidation reactions of N-hydroxylation and N-oxidation that occur at the nitrogen atom include amine or amino group of a foreign compound. N-Hydroxylation involves the substitution of one hydrogen atom in an amino group with a hydroxy group, while N-oxidation involves the addition of oxygen to an amino group. Oxidative enzymes that catalyze oxidation reactions at the nitrogen atom include

CYP450 and flavin-containing monooxygenase. In CYP450-catalyzed oxidation reaction, amine is oxidized to form hydroxyl amine, and imine is oxidized to form oxime in N-hydroxylation, while, in flavin monooxygenase-catalyzed reaction, oxygen atom is added to nitrogen in an amine to form a hydroxyl group.

5.3.3 *Oxidation of Unsaturated Hydrocarbon*

Oxidation of an unsaturated hydrocarbon is an important step in foreign compound metabolism. CYP450-dependent monooxygenase catalyzes oxidation reaction to convert unsaturated aliphatic or aromatic hydrocarbon to form epoxide, where an oxygen atom is inserted into a C=C bond of unsaturated aliphatic or aromatic hydrocarbon to form epoxide. CPY450-produced epoxide can undergo further enzymatic reaction catalyzed by epoxide hydrolase to yield dihydrodiol, while the unhydrolyzed epoxide can either form conjugate with glutathione or react with proteins or DNA.

5.4 Reductive Enzymes

Nitroreductases and azoreductases are two known reductive enzymes for their ability to catalyze the reduction of the nitro groups of a wide range of aromatic substrates and the azo compounds to primary amine metabolites. Nitro compounds are found in chemicals such as industrial solvents, insecticides, and food preservatives.

Azo compounds (Ar–N=N–Ar′) are strongly colored and are widely used as colorants in foods, cosmetics, pharmaceuticals, and textile and printing industries. Aldo-keto reductases catalyze the NAD(P)H-dependent reduction of a large number of carbonyl foreign compounds, such as aldehydes and ketones, to produce primary and secondary alcohols, respectively.

Carbonyl reductase catalyzes the reduction of carbonyls and a wide range of other organic compounds and has a protective role toward xenobiotic carbonyls and quinones. Reactive carbonyls and quinones are able to covalently modify DNA, leading to promote oxidative stress.

5.4.1 *Nitroreductase*

Nitro compounds such as nitro-aromatic and nitro-heterocyclic derivatives constitute a large group of chemicals that are characterized by the presence of nitro ($-NO_2$) groups. Nitroreductase catalyzes the reduction of nitro groups in a wide range of foreign compounds to produce the corresponding hydroxylamines.

In metabolism of nitro compounds, the nitro group is converted to primary amine metabolites, where nitro is initially reduced to nitroso (–NO) and then to hydroxylamine (–NHOH). Hydroxylamine can finally be further converted to primary amine (–NH_2).

Nitroreductases reduce nitro-aromatics to the corresponding nitrosoaromatics and then to the corresponding hydroxyl-amino-aromatics.

5.4.2 *Azoreductase*

Azoreductases are diverse flavoenzymes involved in the biotransformation and detoxification of foreign compounds including azo dyes, nitro-aromatics, and azoic drugs. A typical azo dye contains one or more characteristic azo bonds (–N=N–) in its complex aromatic structure. Azoreductases catalyze the reduction reaction of azo compounds to produce primary amine metabolites. Azo is initially reduced to hydrazo and finally to primary amine.

In azoreductase-catalyzed reactions, two equivalents of NAD(P)H are used to reduce one equivalent of the substrate (azo compound). During the reduction reaction, azo compounds are converted into hydrazo compounds, where the nitrogen-nitrogen double bond is sequentially reduced. Azo compounds are finally cleavage to yield amines.

5.4.3 *Aldo-Keto Reductase*

Aldo-keto reductases (AKRs) belong to a protein superfamily that mainly contains monomeric cytosolic oxidoreductases. AKR superfamily involves biotransformation and detoxification of foreign compounds, such as foods, drugs, and toxins. AKRs catalyze the NAD(P)H-dependent reduction of a wide range of exogenous and endogenous carbonyl substrates, including aliphatic and aromatic aldehydes and ketones. AKRs convert aldehydes and ketones to primary and secondary alcohols. Alcohol and aldehyde dehydrogenases participate in oxidative pathways, whereas reductive routes are accomplished by AKRs.

5.4.4 *Carbonyl Reductase*

Carbonyl reductase is a NADPH-dependent enzyme with substrate specificity for xenobiotic carbonyl compounds. The enzyme plays a critical role in drug metabolism of ketones and aldehydes.

Carbonyl reductase also provides the enzymatic basis of quinone detoxification.

5.5 Reduction at Specific Atom or Group

Foreign compounds containing a nitro or azo group can undergo reduction reactions by interacting with a reducing agent such as NADPH. Reductive reactions catalyzed by azoreductases or nitroreductases occur at nitrogen atom as shown below, while the reduction of carbonyl group to form hydroxyl group is catalyzed by alcohol dehydrogenase.

5.5.1 Reduction at Nitrogen Atom

In azoreductase-catalyzed reduction reaction, N=N in an azo group is reduced to form an amine. While nitroreductase catalyzes the reduction of a nitro ($-NO_2$) group to form an amine.

5.5.2 Reduction of Carbonyl Group

Aldo-keto reductases catalyze the NADPH-dependent reduction of a large number of carbonyl substrates, such as aldehydes and ketones to produce primary and secondary alcohols, respectively. Alcohol dehydrogenase also catalyzes the reduction of carbonyl group to form hydroxyl group.

5.6 Hydrolytic Enzymes

Hydrolytic enzymes (hydrolases) include carboxylesterases and epoxide hydrolases. Hydrolytic enzymes are involved in metabolic activation of various drugs and other xenobiotics.

Carboxylesterases are found in both the endoplasmic reticulum and the cytosol of the cells. Epoxide hydrolases are present in two forms (soluble and microsomal). The enzymes carry out the hydrolysis of epoxides that are usually performed by CYP450-catalyzed reactions.

5.6.1 Carboxylesterase

Carboxylesterases are a serine esterase. The enzymes are composed of a multigene family whose gene products play important roles in metabolic activation and detoxification of various drugs and environmental chemicals. Carboxylesterases catalyze the hydrolysis of a variety of ester- and amide-containing drugs and other chemicals to their respective acids.

Carboxylesterase isozymes consist of carboxylesterase 1 and carboxylesterase 2, which engage in the hydrolysis of esters and amides, especially acting on carboxylic ester bonds.

Carboxylesterases catalyze the hydrolysis reaction of carboxylic ester to form alcohol and carboxylate. The enzymes also catalyze the hydrolysis of carboxylic amide to yield *N*-hydroxide and carboxylate. During carboxylesterase-catalyzed reactions, some foreign compounds are converted to chemically active intermediates, such as the conversion of urea derivative to diazonium hydroxide.

5.6.2 Epoxide Hydrolase

Epoxide hydrolases play an important role in the metabolic activation of xenobiotics such as drugs. The enzymes metabolize foreign compounds that contain an epoxide residue, such as that generated from oxidative activation by CYP450-catalyzed epoxidation. Epoxide hydrolases also convert epoxide residue to two hydroxyl residues through dihydroxylation reaction to form dihydrodiol products. The enzymes also catalyze the trans addition of water to alkene and arene oxides formed by CYP450 bioactivation. Here, epoxide hydrolases are presented as activation enzymes. However, based on their functional properties, sometimes these enzymes are also referred to as detoxification enzymes.

5.7 Hydrolysis at Specific Atom or Group

A substrate containing an ester or amide group can undergo hydrolytic reaction catalyzed by carboxylesterase or hydrolase to produce carboxylic acid and alcohol or carboxylic acid and amine. Enzymatic hydrolysis of amides generally occurs more slowly than that of esters.

5.7.1 Hydrolysis of Ester

Carboxylesterases catalyze hydrolytic reaction converting an ester to carboxylic acid and alcohol.

5.7.2 *Hydrolysis of Amide*

Carboxylesterases also catalyze hydrolysis reaction converting an amide to carboxylic acid and amine.

5.7.3 *Hydrolysis of Epoxide*

Epoxide hydrolases catalyze hydrolysis of epoxide residue to form dihydrodiol.

Bibliography

Abell CW, Kwan SW (2000) Molecular characterization of monoamine oxidases A and B. Prog Nucleic Acid Res Mol Biol 65:129–156

Anzenbacher P, Zanger UM (2012) Metabolism of drugs and other xenobiotics. Wiley VCH, Weinheim

Barski OA, Tipparaju SM, Bhatnagar A (2008) The aldo-keto reductase superfamily and its role in drug metabolism and detoxification. Drug Metab Rev 40(4):553–624

Basaran R, Eke BC (2017) Flavin containing monooxygenases and metabolism of xenobiotics. Turk J Pharm Sci 14(1):90–94

Beedham C (1997) The role of non-P450 enzymes in drug oxidation. Pharm World Sci 19:255–263

Bondy SC, Naderi S (1994) Contribution of hepatic cytochrome P450 systems to the generation of reactive oxygen species. Biochem Pharmacol 48(1):155–159

Brash AR (1999) Lipoxygenases: occurrence, functions, catalysis, and acquisition of substrate. J Biol Chem 274:23679–23682

Klaassen CD (ed) (2018) Casarett & Doull's toxicology: the basic science of poisons, 9th edn. McGraw-Hill, New York

Cashman JR (2002) Human flavin-containing monooxygenase (form 3): polymorphisms and variations in chemical metabolism. Pharmacogenomics 3:325–339

Cashman JR, Zhang J (2006) Human flavin-containing monooxygenases. Annu Rev Pharmacol Toxicol 46:65–100

Chen CH (2012) Activation and detoxification enzymes: functions and implications. Springer, New York

Ross D (2004) Quinone reductases multitasking in the metabolic world. Drug Metab Rev 36 (3–4):639–654

Edmondson DE, Mattevi A, Binda C et al (2004) Structure and mechanism of monoamine oxidase. Curr Med Chem 11:1983–1993

Edenberg HJ (2007) The genetics of alcohol metabolism: role of alcohol dehydrogenase and aldehyde dehydrogenase variants. Alcohol Res Health 30(1):5–13

Gershater MC, Cummins I, Edwards R (2007) Role of a carboxylesterase in herbicide bioactivation in Arabidopsis thaliana. J Biol Chem 282:21460–21466

Guengerich FP (2008) Cytochrome p450 and chemical toxicology. Chem Res Toxicol 21(1):70–83

Guengerich FP (2002) Cytochrome P450. In: Ioannides C (ed) Enzyme system that metabolize drugs and other xenobiotics. Wiley, New York

Hollenberg PF (1992) Mechanisms of cytochrome P450 and peroxidase-catalyzed xenobiotic metabolism. FASEB J 6(2):686–694

Hyndman D, Bauman DR, Heredia VV, Penning TM (2003) The aldo-keto reductase superfamily homepage. Chem Biol Interact 143–144:621–631

Ioannides C (2002) Xenobiotic metabolism: an overview. In: Ioannides C (ed) Enzymes systems that metabolise drugs and other xenobiotics. Wiley, New York

Jin Y, Penning TM (2007) Aldo-keto reductases and bioactivation/detoxification. In: Cho AK, Blaschke TF, Insel PA (eds) Annual review of pharmacology and toxicology, pp 263–292.

Kedderis GL, Hollenberg PF (1983) Characterization of the N-demethylation reactions catalyzed by horseradish peroxidase. J. Biol. Chem 258(13):8129–8138

Krueger SK, Williams DE (2005) Mammalian flavin-containing monooxygenases: structure/function, genetic polymorphisms and role in drug metabolism. Pharmacol Ther 106(3):357–387

Kulkarni AP (2001) Lipoxygenase--a versatile biocatalyst for biotransformation of endobiotics and xenobiotics. Cell Mol Life Sci 58:1805–1825

Hosokawa M (2008) Structure and catalytic properties of carboxylesterase isozymes involved in metabolic activation of prodrugs. Molecules 13(2):412–431

Meijer J, DePierre JW (1988) Cytosolic epoxide hydrolase. Chem Biol Interact 64:207–249

O'Brien PJ (2000) Peroxidases. Chem Biol Interact 129:113–139

Misal SA, Gawai KR (2018) Azoreductase: a key player of xenobiotic metabolism. Bioresour Bioprocess 5:17

Nassar AF (ed) (2010) Biotransformation and metabolite elucidation of xenobiotics. Wiley, Hoboken

Oppermann U (2007) Carbonyl reductases: the complex relationships of mammalian carbonyl- and quinone-reducing enzymes and their role in physiology. Annu Rev Pharmacol Toxicol 47:293–322

Oppermann UC, Maser E (2000) Molecular and structural aspects of xenobiotic carbonyl metabolizing enzymes. Role of reductases and dehydrogenases in xenobiotic phase I reactions. Toxicology 144(1–3):71–81

Parkinson A, Ogilvie BW (2008) Biotransformation of xenobiotics. In: Klaassen CD (ed) Casarett and Doull's toxicology: the basic science of poisons. McGrawHill, New York

Satoh T, Hosokawa M (2006) Structure, function and regulation of carboxylesterases. Chem Biol Interact 162:195–211

Satoh T, Hosokawa M (1998) The mammalian carboxylesterases: from molecules to functions. Annu Rev Pharmacol Toxicol 38:257–288

Strolin Benedetti M, Tipton KF (1998) Monoamine oxidases and related amine oxidases as phase I enzymes in the metabolism of xenobiotics. J Neural Transm Suppl 52:149–171

Thomas PE, Ryan DE, Von Bahr C, Glaumann H, Levin W (1982) Human liver microsomal epoxide hydrolase. Correlation of immunochemical quantitation with catalytic activity. Mol Pharmacol 37:190–195

Tipton KF, Benedetti MS (2002) Amine oxidases and the metabolism of xenobiotics. In: Ioannides C (ed) Enzyme system that metabolize drugs and other xenobiotics. Wiley, New York

Wermuth B, Platts KL, Seidel A, Oesch F (1986) Carbonyl reductase provides the enzymatic basis of quinone detoxification in man. Biochem Pharmacol 35(8):1277–1282

Ziegler DM (2002) An overview of the mechanism, substrate specificities, and structure of FMOs. Drug Metab Rev 34:503–511

Chapter 6
Functionalization Reactions Catalyzed by Activation Enzymes

Activation enzymes catalyze oxidation, reduction, and hydrolysis reactions, resulting in the introduction of functional groups to lipophilic foreign compounds, thus increasing the water solubility of parent compounds and facilitating their excretion from the body. The functional reactions introduced to foreign compounds include N-oxidation, S-oxidation, aromatic and aliphatic hydroxylation, O- and N-dealkylation, hydrolysis, and epoxidation.

In continuation of the previous chapter that describes the functional properties of oxidases, reductases, and hydrolases as well as the atoms and groups that are involved in oxidation, reduction, and hydrolysis reactions, the present chapter discusses functionalization reactions catalyzed by major activation enzymes, including cytochrome P450 oxidase, flavin monooxygenase, amine oxidase, nitroreductase, azoreductase, molybdenum hydroxylase, epoxide hydrolase, alcohol dehydrogenase, and aldehyde dehydrogenase.

6.1 Activation Metabolism

Activation enzyme-catalyzed chemical reactions presented below are focused on the introduction of functional group to foreign compound (functionalization), where the equation may or may not be balanced.

The original version of this chapter was revised. The correction to this chapter is available at https://doi.org/10.1007/978-3-030-41679-9_20

C.-H. Chen, *Xenobiotic Metabolic Enzymes: Bioactivation and Antioxidant Defense*, https://doi.org/10.1007/978-3-030-41679-9_6

6.1.1 Cytochrome P450 Oxidase

Cytochrome P450 oxidase family is a monooxygenase that plays a prominent role in the metabolism of a variety of foreign compounds, including drugs, foods, and environmental and industrial chemicals. In addition to carbon hydroxylation in the metabolism of sterols and alkanes, oxidation reactions catalyzed by CYP450s include hydroxylation, epoxidation, dehydrogenation, heteroatom oxygenation, heteroatom dealkylation, and oxidation of aromatic rings.

(a) Hydroxylation of aliphatic or aromatic compound
Hydroxylation occurring in aliphatic and aromatic compounds involves the oxidation of hydrogen in an alkyl or aromatic group, resulting in the formation of alcohol. Hydroxylation is a common process in the metabolism of sterols and alkanes. Hydroxylation reaction requires NADPH and oxygen:

$$R - CH_3 + O_2 + NADPH + H^+ \rightarrow R - CH_2OH + H_2O + NADP^+$$
$$\Phi - H_3 + O_2 + NADPH + H^+ \rightarrow \Phi - OH + H_2O + NADP^+$$

where R and Φ denote aliphatic and aromatic groups, respectively.

(b) Epoxidation of ether
An epoxide is a cyclic ether with three members ring, which arises from oxidative metabolism of foreign compounds via enzymatic oxidation reactions. Formation of epoxide occurs as CYP450-mediated metabolites of unsaturated carbon-carbon bonds. The resultant epoxides are unstable and reactive:

$$R_1C{=}CR_2 + O_2 + NADPH + H^+ \rightarrow R_1\overset{O}{\overbrace{C{-}C}}R_2 + H_2O + NADP^+$$

where R_1 and R_2 represent alkyl groups.

(c) Dehydrogenation of alcohol or aldehyde
CYP450-catalyzed dehydrogenation reaction results in oxidizing alcohol to form aldehyde or aldehyde to form carboxylic acid:

$$R - CH_2OH + O_2 + NADPH + H^+ \rightarrow RCH = O + 2H_2O + NADP^+$$
$$R - CHO + O_2 + NADPH + H^+ \rightarrow RCOOH + H_2O + NADP^+$$

where R denotes an alkyl group.

(d) Oxidation of N- or S-compound
Oxygenation reactions that involve N- or S-compounds apply to amines and sulfides. Amines are oxidized to hydroxyl amines, and thioethers are oxidized to sulfoxides:

$$\Phi - NH_2 + O_2 + NADPH + H^+ \rightarrow \Phi - NHOH + H_2O + NADP^+$$
$$\Phi - S - CH_3 + O_2 + NADPH + H^+ \rightarrow \Phi - SO - CH_3 + H_2O + NADP^+$$

where Φ represents an aromatic group.

(e) Dealkylation of ether, amide, or carboxylic acid
The cleavage of ether, amide, or carboxylic acid is a common CYP450-catalyzed reaction:

$$R - O - R' + O_2 + NADPH + H^+ \rightarrow RCHO + HOR' + H_2O + NADP^+$$

$$\begin{array}{l} R{-}C{=}O + O_2 + NADPH + H^+ \rightarrow RCHO + NH_3OH + H_2O + NADP^+ \\ \quad | \\ \quad NH_2 \end{array}$$

$$\begin{array}{l} R{-}C{=}O + O_2 + NADPH + H^+ \rightarrow HCHO + HCOOH + H_2O + NADP^+ \\ \quad | \\ \quad OH \end{array}$$

(f) Oxidation of carbon on aromatic ring
CYP450s catalyze the oxidation of a carbon on an aromatic ring:

$$\Phi - X + O_2 + NADPH + H^+ \rightarrow \Phi - O - X + NADP^+ + H_2O$$

where X represents H, N, or S attaching to a carbon on the aromatic ring.

6.1.2 *Flavin Monooxygenase*

Flavin-containing monooxygenase also belongs to oxidative enzymes. These enzymes exhibit functions that overlap with CYP450 oxidase. Monooxygenases are also involved in the oxidation of numerous organic compounds containing nitrogen, sulfur, or phosphorus to form nitrogen, sulfur, or phosphorus oxide.

Unlike CYP450s, flavin monooxygenase-catalyzed reactions utilize flavin adenosine dinucleotide (FAD) as the coenzyme. Flavin monooxygenase-catalyzed fatty acid oxidation reaction can be represented as:

$$R - CH_2 - CH_2 - CH_2 - CO - S - CoA + FAD \rightarrow R - CH_2 - CH = CH - CO - S - CoA + FADH_2$$

where $CH_2{-}CH_2$ bond is oxidized to form CH=CH.

6.1.3 *Amine Oxidase*

Amine oxidase catalyzes the oxidation of amines in the metabolism of foreign compounds. Amino oxidase-catalyzed reaction involves the oxidative cleavage of the α-H in aliphatic or aromatic amines:

$$RCH_2NR'R'' + H_2O + O_2 \rightarrow RCHO + NHR'R'' + H_2O_2$$

The products of amine oxidase-catalyzed reactions such as aldehyde, ammonia, and hydrogen peroxide can be potentially toxic. The resulting hydrogen peroxide is the source of hydroxyl radical (·OH).

The produced aldehyde may be further metabolized by aldehyde oxidase or aldehyde reductase to form carboxylic acid or alcohol:

$$RCHO + NAD + H_2O \rightarrow RCOOH + NADH_2$$

$$RCHO + NADH_2 \rightarrow RCH_2OH + NAD$$

6.1.4 Aldehyde Oxidase and Xanthine Oxidase

Aldehyde oxidase and xanthine oxidase, among the members of molybdenum hydroxylase family, are important in the metabolism of drugs and other xenobiotics. Molybdenum hydroxylases exhibit oxidase activity toward a variety of heterocyclic compounds and aldehydes.

Aldehyde oxidase-catalyzed oxidation reaction inserts the oxygen atom into the substrate (foreign compound), such as the conversion of an aldehyde to form an acid and hydrogen peroxide, where oxygen is from water instead of molecular oxygen:

$$HCHO + H_2O + O_2 \rightleftharpoons HCOOH + H_2O_2 + H^+$$

Xanthine oxidase catalyzes the oxidation of hypoxanthine to xanthine, which is further catalyzed to uric acid. Xanthine oxidase also plays an important role in the catabolism of purines:

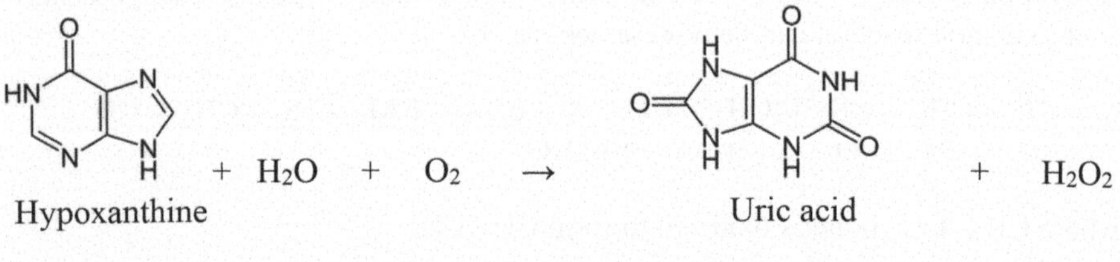

6.1.5 Peroxidase

Peroxidases are a large family of enzymes that catalyze the conversion of peroxides to form alcohols:

$$RO - OR' + 2H^+ \rightarrow ROH + R'OH$$

where RO–OR′ are such as hydrogen peroxide or organic hydroperoxides.

6.1.6 *Nitroreductase*

Nitroreductase catalyzes the reduction of nitro compounds to form primary amine metabolites:

$$RNO_2 + NADPH + H^+ \rightarrow RNO + NADP^+ + H_2O$$

$$RNO + NADPH + H^+ \rightarrow RNHOH + NADP^+$$

$$RNHOH + NADPH + H^+ \rightarrow RNH_2 + NADP^+ + H_2O$$

where the nitro group ($-NO_2$) is initially reduced to nitroso (–NO), then to hydroxylamine (–NHOH), and finally to primary amine ($-NH_2$).

6.1.7 *Azoreductase*

Azoreductase catalyze the reduction of azo compounds, where azo is initially reduced to hydrazo and finally to primary amine:

$$\Phi - N = N - \Phi' + NADPH + H^+ \rightarrow \Phi - NH - NH - \Phi' + NADP^+$$

$$\Phi - NH - NH - \Phi' + NADPH + H^+ \rightarrow \Phi - NH_2 + \Phi' - NH_2 + NADP^+$$

where Φ denotes an aromatic group, which may be replaced with aliphatic group.

6.1.8 *Carbonyl Reductase*

Carbonyl reductase is an enzyme with substrate specificity for xenobiotic carbonyl compounds. Carbonyl reductase catalyzes the reduction of a variety of xenobiotic quinones derived from polycyclic aromatic hydrocarbons:

$$R - CO - R' + NADPH + H^+ \rightarrow R - CHOH - R' + NADP^+$$

$$NAD(P)H + H^+ + quinone \rightarrow NAD(P)^+ + hydroquinone$$

6.1.9 *Alcohol Dehydrogenase*

Alcohol dehydrogenase catalyzes the conversion of primary or secondary alcohol to aldehyde or ketone. This reaction requires NAD as a coenzyme:

$$RCH_2OH + NAD^+ \rightarrow RCHO + NADH + H^+$$

The produced aldehyde is usually toxic, which can be further oxidized to acid.

6.1.10 *Aldehyde Dehydrogenase*

Aldehyde dehydrogenases are a group of enzymes that catalyze the oxidation of aldehydes. The oxidation reaction converts aldehydes to carboxylic acids. The oxygen comes from a water molecule:

$$R - CHO + O_2 \rightarrow R - COOH$$

6.1.11 *Carboxylesterase*

Carboxylesterase catalyzes the hydrolysis reaction specifically involving the carboxylic ester bond:

$$R - COOR' + H_2O \rightarrow R'OH + RCOO^- + H^+$$

where $R{-}COOR'$ represents a carboxylic ester. The reaction products are alcohol and carboxylate.

6.2 Oxidation Reactions

Metabolic enzyme-catalyzed oxidation reactions involving carbon and nitrogen atoms and unsaturated hydrocarbon of foreign compounds are presented below.

6.2.1 Oxidation at Carbon Atom

Oxidation reaction at carbon atom occurs on an aromatic ring, alcohol, aldehyde, or other functional group of a foreign compound. Typical reactions of oxidation at carbon atom catalyzed by CYP450 oxidase, alcohol dehydrogenase, aldehyde dehydrogenases, xanthine oxidases, and peroxidases are shown below.

(a) Cytochrome P450 oxidase
An oxygen atom is inserted in a C–H bond of an aromatic compound to form phenol:

$$\Phi - H + O_2 + NADPH + H^+ \rightarrow \Phi - OH + H_2O + NADP^+$$

(b) Alcohol dehydrogenase
Oxidation reaction converts aliphatic, aromatic, or cyclic alcohol to aldehyde by removing two hydrogens (one attached to carbon and another attached to oxygen):

$$R - CH_2OH + NAD^+ \rightarrow R - CHO + NADH + H^+$$

Further reaction oxidizes aldehyde to form carboxylic acid:

$$R - CHO + H_2O + NAD^+ \rightarrow R - COOH + NADH + H^+$$

(c) Aldehyde oxidase
Aldehyde oxidase catalyzes the oxidation of oxygen atom in C–H bond of aldehyde to yield carboxylic acid:

$$R - CHO + H_2O + O_2 \rightarrow R - COOH + H_2O_2$$

(d) Peroxidase
Peroxidase catalyzes the oxidation of alcohol to form aldehyde:

$$R - CH_2OH + O_2 \rightarrow R - CHO + H_2O$$

6.2.2 *Oxidation at Nitrogen Atom*

Oxidation reactions taking place at a nitrogen atom of foreign compound include N-hydroxylation and N-oxidation, which occur in amine or amino group of a foreign compound. N-Hydroxylation consists of the substitution of one hydrogen atom in an amino group with a hydroxy group, while N-oxidation involves the addition of oxygen to an amino group. Oxidative enzymes that catalyze oxidation reactions at the nitrogen atom include CYP450 oxidase and flavin-containing monooxygenase.

(a) Cytochrome P450 oxidase
Amine group of a foreign compound is oxidized to form hydroxyl amine:

$$2\ \Phi\text{-}\underset{\substack{|\\ \text{H}}}{\text{N}}\text{-CO-CH}_3 + \text{O}_2 \rightarrow 2\ \Phi\text{-}\underset{\substack{|\\ \text{OH}}}{\text{N}}\text{-CO-CH}_3$$

Imine group is oxidized to form oxime in N-hydroxylation:

$$2\ \text{R-}\underset{\substack{|\\ \text{R'}}}{\text{C}}\text{=N-H} + \text{O}_2 \rightarrow 2\ \text{R-}\underset{\substack{|\\ \text{R'}}}{\text{C}}\text{=N-OH}$$

$$2\ \Phi\text{-}\underset{\substack{|\\ \text{H}}}{\text{C}}\text{=N-H} + \text{O}_2 \rightarrow 2\ \Phi\text{-}\underset{\substack{|\\ \text{H}}}{\text{C}}\text{=N-OH}$$

(b) Flavin monooxygenases
An oxygen atom is added to nitrogen in an amine of foreign compound to form a hydroxyl group:

$$2\ \Phi\text{-}\overset{\substack{\text{CH}_3\\ |}}{\text{N}}\text{-H} + \text{O}_2 \rightarrow 2\ \Phi\text{-}\overset{\substack{\text{CH}_3\\ |}}{\text{N}}\text{-OH}$$

6.2.3 *Oxidation of Unsaturated Hydrocarbon*

Oxidation of an unsaturated hydrocarbon is an important step in foreign compound metabolism. CYP450-dependent monooxygenase catalyzes oxidation reaction to convert unsaturated aliphatic or aromatic hydrocarbon to form epoxide, where an oxygen atom is inserted into a C=C bond of unsaturated aliphatic or aromatic hydrocarbon to form epoxide:

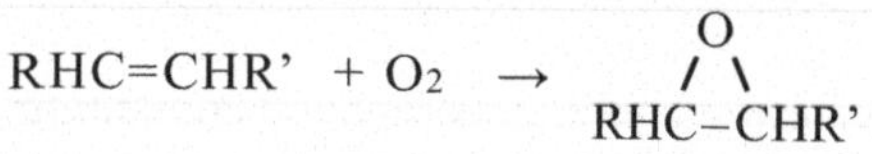

The above produced epoxide can undergo further enzymatic reaction catalyzed by epoxide hydrolase to yield dihydrodiol, as shown in the following reaction, while the unhydrolyzed epoxide can either form conjugate with glutathione or react with proteins or DNA:

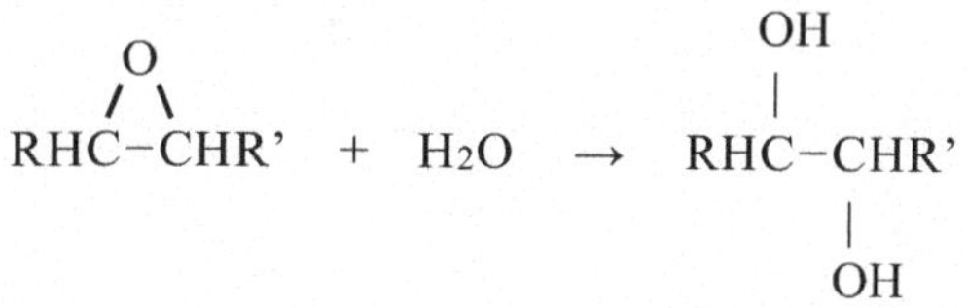

6.3 Reduction Reactions

Nitroreductase and azoreductase catalyze the reduction of nitro- and azo-containing foreign compounds to primary amine metabolites. Nitroreductase reduces nitro-aromatics to nitrosoaromatics and then to hydroxylamino-aromatics. Azoreductase, a diverse flavoenzyme, catalyzes the reduction reaction of azo compounds to primary amine metabolites.

6.3.1 Reduction at Nitrogen Atom

(a) Nitroreductase

A nitro ($-NO_2$) group of a foreign compound is reduced to form an amine or hydroxylamine catalyzed by nitroreductase. Hydroxylamine can be further converted to primary amine ($-NH_2$):

$$R - \Phi - NO_2 + 2NADPH + 2H^+ \rightarrow R - \Phi - NH_2 + H_2O + 2NADP^+$$

(b) Azoreductase

Azoreductase catalyzes the reduction of N=N in an azo group of a foreign compound to form an amine:

$$\Phi - N = N = \Phi' + 2NADPH + 2H^+ \rightarrow \Phi - NH_2 + H_2N - \Phi' + 2NADP^+$$

6.3.2 *Reduction of Carbonyl Group*

Aldo-keto reductases catalyze a foreign compound containing aldehyde or ketone to produce primary or secondary alcohol:

$$
\begin{array}{ccccccc}
 & & & & \text{OH} & & \\
 & & & & | & & \\
\text{R–C=O} & + & \text{NADPH} & \rightarrow & \text{R–C–OH} & + & \text{NADP}^{+} \\
| & & & & | & & \\
\text{R'} & & & & \text{R'} & &
\end{array}
$$

6.4 Hydrolysis Reactions

Hydrolase enzymes including carboxylesterase and epoxide hydrolase catalyze hydrolytic reactions of various drugs and other xenobiotics.

6.4.1 *Hydrolysis of Ester and Amide*

Carboxylesterase catalyzes the hydrolysis of a variety of esters and amides. Hydrolysis reaction of carboxylic ester forms alcohol and carboxylate, and the hydrolysis of carboxylic amide yields N-hydroxide and carboxylate.

Carboxylesterase catalyzes hydrolytic reaction of an ester to yield carboxylic acid and alcohol. The enzyme also catalyzes hydrolysis reaction by converting an amide to carboxylic acid and amine:

$$
\begin{array}{ccccccc}
\text{O=C–OC}_2\text{H}_5 & & & & \text{COOH} & & \\
| & & & & | & & \\
\text{RHCHR'} & + & \text{H}_2\text{O} & \rightarrow & \text{RHCHR'} & + & \text{C}_2\text{H}_5\text{OH}
\end{array}
$$

$$
\begin{array}{ccccccc}
\text{R} & & & & & & \\
| & & & & & & \\
\Phi\text{-CO-N-H} & + & \text{H}_2\text{O} & \rightarrow & \Phi\text{-COOH} & + & \text{H}_2\text{N-R}
\end{array}
$$

6.4.2 *Hydrolysis of Epoxide*

Epoxide hydrolase catalyzes the conversion of epoxide residue to hydroxyl residue and also catalyzes hydroxylation reaction to form dihydrodiol product:

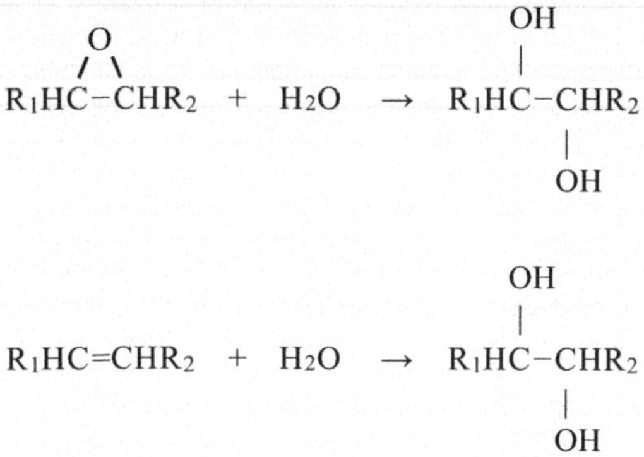

Bibliography

Beedham C (1985) Molybdenum hydroxylases as drug-metabolizing enzymes. Drug Metab Rev 16 (1- 2):119–156

Beedham C (1998) Molybdenum hydroxylases. In: Gorrod JW, Oeschlager H, Caldwell J (eds) Metabolism of xenobiotics. Taylor and Francis, London, pp 51–58

Benedetti MS (2001) Biotransformation of xenobiotics by amine oxidases. Fundam Clin Pharmacol 15(2):75–84

Benedetti MS, Dostert P (1994) Contribution of amine oxidases to the metabolism of xenobiotics. Drug Metab Rev 26(3):507–535

Cashman JR (2000) Human flavin-containing monooxygenase: substrate specificity and role in drug metabolism. Curr Drug Metab 1(2):181–191

Cashman JR, Zhang J (2006) Human flavin-containing monooxygenases. Annu Rev Pharmacol Toxicol 46:65–100

Chen CH (2012) Activation and detoxification enzymes: functions and implications. Springer, New York

Christofferson A, Wilkie J (2009) Mechanism of CB1954 reduction by Escherichia coli nitroreductase. Biochem Soc Trans 37(pt 2):413–418

Dawson JH (1998) Probing structure-function relations in heme-containing oxygenases and peroxidases. Science 240:433–439

Edmondson DE, Mattevi A, Binda C, Li M, Hubálek F (2004) Structure and mechanism of monoamine oxidase. Curr Med Chem 11(15):1983–1993

Forrest GL, Gonzalez B (2000) Carbonyl reductase. Chem Biol Interact 129(1–2):21–40

Guengerich FP (1991a) Reactions and significance of cytochrome P-450 enzymes. J Biol Chem 266 (16):10019–10022

Guengerich FP (1996) The chemistry of cytochrome P450 reactions. In: Ioannides C (ed) Cytochrome P450: metabolic and toxicological aspects. CRC Press, Boca Raton, pp 55–74

Guengerich FP, Peterson LA, Böcker RH (1988) Cytochrome P-450-catalyzed hydroxylation and carboxylic acid ester cleavage of Hantzsch pyridine esters. J Biol Chem 263(17):8176–8183

Guengerich FP, Shimada T (1991b) Oxidation of toxic and carcinogenic chemicals by human cytochrome P-450 enzymes. Chem Res Toxicol 4(4):391–407

Heikkila RE, Kindt MV, Sonsalla PK, Giovanni A, Youngster SK, McKeown KA, Singer TP (1988) Importance of monoamine oxidase A in the bioactivation of neurotoxic analogs of 1-methyl-4-phenyl- 1,2,3,6-tetrahydropyridine. Proc Natl Acad Sci 85(16):6172–6176
Hodgson E, Levi PE (1992) The role of the flavin-containing monooxygenase (EC 1.14.13.8) in the metabolism and mode of action of agricultural chemicals. Xenobiotica 22(9–10):1175–1183
Ioannides C (ed) (2002) Enzyme systems that metabolise drugs and other xenobiotics. Wiley, New York
Ito K, Nakanishi M, Lee WC, Zhi Y, Sasaki H, Zenno S, Saigo K, Kitade Y, Tanokura M (2008) Expansion of substrate specificity and catalytic mechanism of azoreductase by X-ray crystallography and site-directed mutagenesis. J Biol Chem 283(20):13889–13896
Koder RL, Haynes CA, Rodgers ME, Rodgers DW, Miller AF (2002) Flavin thermodynamics explain the oxygen insensitivity of enteric nitroreductases. Biochemistry 41(48):14197–14205
Krenitsky TA, Neil SM, Elion GB, Hitchings GH (1972) A comparison of the specificities of xanthine oxidase and aldehyde oxidase. Arch Biochem Biophys 150(2):585–599
Kulkarni AP, Hodgson E (1984) The metabolism of insecticides: the role of monooxygenase enzymes. Annu Rev Pharmacol Toxicol 24:19–42
Li R, Bianchet MA, Talalay P, Amzel LM (1995) The three-dimensional structure of NAD(P)H: quinone reductase, a flavoprotein involved in cancer chemoprotection and chemotherapy: mechanism of the two-electron reduction. Proc Natl Acad Sci U S A 92(19):8846–8850
Meunier B, de Visser SP, Shaik S (2004) Mechanism of oxidation reactions catalyzed by cytochrome p450 enzymes. Chem Rev 104(9):3947–3980
de Montellano O, Paul R, de Montellano O (2005) Cytochrome P450: structure, mechanism, and biochemistry, 3rd edn. Kluwer Academic/Plenum Publishers, New York
O'Brien PJ (2000) Peroxidases. Chem Biol Interact 129:113–139
Parkinson A, Ogilvie BW (2008) Biotransformation of xenobiotics. In: Klaassen CD (ed) Casarett and Doull's toxicology: the basic science of poisons. McGrawHill, New York
Rajagopalan KV (1980) Xanthine oxidase and aldehyde oxidase. In: Jakoby WB (ed) Enzymatic basis of detoxification. Academic, New York
Rafii F, Cerniglia CE (1995) Reduction of azo dyes and nitroaromatic compounds by bacterial enzymes from the human intestinal tract. Environ Health Perspect 103(suppl 5):17–19
Satoh T, Hosokawa M (2006) Structure, function and regulation of carboxylesterases. Chem Biol Interact 162(3):195–211
Satoh T, Hosokawa M (1998) The mammalian carboxylesterases: from molecules to functions. Annu Rev Pharmacol Toxicol 38:257–288
Seitz HK, Oneta CM (1998) Gastrointestinal alcohol dehydrogenase. Nutr Rev 56:52–60
Senter PD, Marquardt H, Thomas BA, Hammock BD, Frank IS, Svensson HP (1996) The role of rat serum carboxylesterase in the activation of paclitaxel and camptothecin prodrugs. Cancer Res 56(7):1471–1474
Shimada T, Martin MV, Pruess-Schwartz D, Marnett LJ, Guengerich FP (1989) Roles of individual human cytochrome P-450 enzymes in the bioactivation of benzo(a)pyrene, 7,8-dihydroxy-7,8-dihydrobenzo(a)pyrene, and other dihydrodiol derivatives of polycyclic aromatic hydrocarbons. Cancer Res 49(22):6304–6312
Strolin Benedetti M, Tipton KF (1998) Monoamine oxidases and related amine oxidases as phase I enzymes in the metabolism of xenobiotics. J Neural Transm Suppl 52:149–171
Uetrecht J (2003) Bioactivation. In: Lee JS, Obach S, Fisher MB (eds) Drug metabolizing enzymes. Marcel Dekker, New York, Chap 3
Vasiliou V, Pappa A, Estey T (2004) Role of human aldehyde dehydrogenases in endobiotic and xenobiotic metabolism. Drug Metab Rev 36(2):279–299
Wong LL (1998) Cytochrome P450 monooxygenases. Curr Opin Chem Biol 2:263–268
Ziegler DM (1998) Flavin-containing monooxygenases: catalytic mechanism and substrate specificities. Drug Metab Rev 19(1):1–32

Chapter 7
Detoxifying Metabolism: Detoxification Enzymes

Activation enzymes catalyze functionalization reactions that incorporate a functional group to a foreign compound, resulting in the formation of an intermediate or metabolite. Many of intermediates or metabolites contain highly reactive chemical groups, which have the potential to react with cellular components, leading to adverse health effects. Following activation enzyme-catalyzed functionalization reactions, the generated intermediates or metabolites undergo detoxification metabolism to form more hydrophilic and less reactive metabolic compounds, thus facilitating their excretion from the body. Unlike activation enzymes that catalyze functionalization reactions to increase water solubility of foreign compounds, detoxification enzymes catalyze conjugation reactions that deactivate and detoxify foreign compounds.

The entering of foreign compounds into the cells is referred to as phase 0 metabolism, and phases I and II are referred to activation metabolism and detoxification metabolism, respectively. Following conjugation metabolism, the produced conjugates are transported to the kidneys and are available for excretion in the urine or through the apical surface of the liver hepatocytes into the bile ducts. The excretion of conjugated foreign compounds from the cells, mediated by ATP-dependent export pumps such as multidrug resistance proteins, is referred to as phase III metabolism. Schematic representations of phases 0, I, II, and III metabolisms are shown in Fig. 7.1.

7.1 Excretors

Detoxifying metabolism conjugates a functionalized foreign compound with a small, endogenous molecule (conjugating ligand), resulting in reducing the reactivity of intermediate or metabolite, diminishing its potential toxicity to the cells, and also increasing its water solubility, thus facilitating its excretion from the body. Detoxification enzymes are therefore referred to as excretors.

C.-H. Chen, *Xenobiotic Metabolic Enzymes: Bioactivation and Antioxidant Defense*,
https://doi.org/10.1007/978-3-030-41679-9_7

Fig. 7.1 Metabolic phases prior to the excretion of foreign compounds

Foreign compounds
Phase 0 metabolism ↓
Entering into the cells
Phase I metabolism ↓ Functionalization
Intermediates or metabolites
Phase II metabolism ↓ Conjugation
Conjugated xenobiotics
Phase III metabolism ↓ ATP-dependent exporter
Export from the cells
Urine ↓ bile
Excretion from the body

Although conjugation reactions generally result in an increase in the solubility of foreign compounds, there are exceptional cases, such as methylation and acetylation that produce conjugates with less solubility in water. Moreover, the efficiency of the detoxification process depends on not only the functions of activation metabolism but also detoxification metabolism.

Detoxification enzymes mainly belong to the class of transferases, which are responsible for catalytic conjugation reactions in detoxifying metabolism. Besides transferases, a few non-conjugation enzymes that also engage in detoxifying metabolism are also classified as phase II detoxification enzymes.

7.2 Conjugation Enzymes

Detoxification enzymes that catalyze conjugation reactions are mainly composed of a set of transferase enzymes, including uridine diphosphate-glucuronosyltransferases, glutathione *S*-transferases, sulfotransferases, aceyltransferases, methyltransferases, and acyltransferases. Among them, uridine diphosphate-glucuronosyltransferases, glutathione *S*-transferases, and sulfotransferases are major detoxification enzymes. Table 7.1 presents detoxification enzymes, conjugation cofactors, and functional groups of foreign compounds that are involved in conjugation reactions. The functional properties of these conjugation enzymes are described below.

7.2.1 Uridine Diphosphate-Glucuronosyltransferase

Uridine diphosphate-glucuronosyltransferase (UGT) is one of the most important transferases that catalyze conjugation reactions in detoxifying metabolism. Glucuronidation accounts for about 35% of drug conjugations. The enzyme is located

Table 7.1 Detoxification enzymes, conjugation cofactors, and functional groups

Detoxification enzyme	Conjugation cofactor	Functional groups
(a) Conjugation enzymes		
Uridine diphosphate-glucuronosyltransferases (UGTs)	Glucuronic acid	Phenol, sulfhydryl, carboxylic, amine
Glutathione *S*-transferases (GST)	Glutathione	Halide, carbonyl or epoxide
Sulfotransferases (ST)	Sulfonate	Hydroxyl, amine
N-Acetyltransferases (NAT)	Acetyl group	Amine or hydrazine
Methyltransferases	Methyl group	Phenol, amine or sulfhydryl
Acyltransferases	Acyl group	Carboxylic, amino acid
(b) Non-conjugation enzymes		
Quinone reductases		Quinones
Epoxide hydrolases		Epoxides

in the endoplasmic reticulum and contains in regions of membrane association and transmembrane domain. UGT activity is therefore highly dependent on lipids. The liver is the major site of glucuronidation in the living organism. Other tissues where glucuronidation occurs include the kidney, gastrointestinal tract, and lungs.

UGTs are a family of enzymes that catalyze the covalent binding of glucuronic acid to a wide range of lipophilic foreign compounds. Glucuronidation reaction also involves the transfer of glucuronic acid from uridine diphosphate (UDP)-glucuronic acid to a functional group of foreign compounds, such as oxygen, nitrogen, sulfur, or carboxyl groups.

As a major pathway in detoxifying metabolism, glucuronidation represents one of the most important detoxifying reactions involving the conversion of lipophilic xenobiotics and their metabolites into hydrophilic conjugates, thus facilitating their excretion from the body. The site of glucuronidation is generally at electron-rich nucleophilic atom (e.g., O, N, or S).

The substrates for glucuronidation are such as alcohols, phenols, carboxylic acids, sulfhydryl, or aromatic and aliphatic amines. Glucuronidation is therefore the primary metabolic reaction for many compounds that contain functional groups such as –OH, –COOH, –SH, and $-NH_2$. Glucuronic acid conjugates with phenolic, carboxylic, or sulfhydryl group of a substrate to form *O*-glucuronide or with an amino group of a substrate to form *N*-glucuronide.

Typical UGT conjugation reactions with glucuronic acid (GA) are shown below.

Oxygen atom of hydroxyl group in phenol is conjugated with glucuronic acid (GA) to form a glucuronide derivative:

$$\Phi - \mathrm{OH} + \mathrm{UDP} - \mathrm{GA} \rightarrow \Phi - \mathrm{O} - \mathrm{GA} + \mathrm{UDP}$$

GA attaches to N atom of amine to form a glucuronide conjugate:

$$\begin{array}{ccccccc} \Phi\text{-N-OH} & + & \text{UDP–GA} & \rightarrow & \text{HO-}\Phi\text{–N} & + & \text{UDP} \\ | & & & & | & & \\ \text{H} & & & & \text{GA} & & \end{array}$$

GA binds to the S atom of thio compound to form a glucuronide conjugate:

$$\mathrm{R-SH+UDP-GA\rightarrow R-S-GA+UDP}$$

GA reacts with carboxylic acid to form a glucuronide conjugate:

$$\Phi-\mathrm{COOH+UDP-GA}\rightarrow\Phi-\mathrm{COOGA+UDP}$$

Detoxification enzymes catalyze GA conjugation reactions that deactivate and detoxify foreign compounds. Meanwhile, glucuronidation significantly increases the aqueous solubility of foreign compounds, making them more readily available for excretion. Although glucuronidation of foreign compounds is considered to be beneficial, however, in some cases, glucuronidation reaction may produce potentially toxic conjugate.

7.2.2 Glutathione S-Transferase

Glutathione *S*-transferases (GSTs) are a multigene family of enzymes that are involved in the metabolism of a wide range of foreign compounds, including arene oxides, unsaturated carbonyls, and organic halides. GST families consist of cytosolic, mitochondrial, and microsomal GSTs, which are differentiated into cytosolic enzymes and membrane-bound enzymes. Glutathione conjugation is a highly important mechanism in metabolizing foreign compounds.

GSTs catalyze the conjugation of reduced glutathione (GSH) with a foreign compound through the formation of a thioether bond between the sulfur atom of GSH and carbon or nitrogen in a foreign compound. The conjugate is subsequently metabolized to form cysteine or *N*-acetylcysteine characteristics.

Foreign compounds that undergo GST-catalyzed conjugation reactions include alkyl and aryl halides, isothiocyanates, α,β-unsaturated carbonyls, and epoxides. GSTs catalyze the conjugation of GSH to foreign compounds via a sulfhydryl group to electrophilic centers on the substrates. In epoxide, reduced GSH binding to C atom of epoxide forms a glutathione conjugate:

```
      O                         OH
     / \                         |
  Φ–C–C–R'  +    GSH     →    R–C–C–R'
     |  |                        |  |
     H  H                        H  SG
```

GSTs detoxify not only electrophiles derived from activation metabolism of xenobiotics but also endogenous metabolites. Electrophiles are potentially toxic species, since they can bind to nucleophilic residues in proteins and nucleic acids, leading to cellular damage and genetic mutation. Moreover, electrophilic metabolites reacting with oxygen are usually the consequence of free radicals that can cause damages to cellular components.

GST-catalyzed conjugation reactions inactivate reactive metabolites or intermediates and also make foreign compounds more water soluble, thus facilitating their excretion from the body. However, in certain cases, GSH conjugation may become an activation step, due to the formation of degradation of metabolites.

Besides serving as a catalyst for conjugation reactions, GST also involves the transfer of glutathione, a reaction that occurs in the cytosol of the liver. The tripeptide glutathione is synthesized from γ-glutamic acid, cysteine, and glycine. GSH has the ability to bind many hydrophobic substances, such as bilirubin, steroids, and polycyclic aromatic hydrocarbons.

7.2.3 Sulfotransferase

Sulfotransferase (ST)-catalyzed conjugation reaction is another major pathway of detoxifying metabolism of foreign compounds. Cytosol STs catalyze sulfonate conjugation reactions of relatively small lipophilic xenobiotics. At least 44 cytosolic STs, constituting of five different families, have been identified. 3′-Phosphoadenosine 5-′-phosphosulfate (PAPS) is the cofactor of STs for the sulfonation reaction.

PAPS is synthesized in tissues to make available an activated form of sulfonate for sulfotransferase-catalyzed reactions. The synthesis is dependent on the availability of sulfonate and the activity of the enzymes involved. A relatively low concentration of PAPS can limit the capacity for sulfonation of foreign compounds.

Sulfonation reaction involves the transfer of a sulfonate group (SO_3^-) from the donor, PAPS, to the acceptor, a nucleophilic group of foreign compounds. In sulfonation conjugation reaction, the nucleophilic attack of oxygen or nitrogen atom in a foreign compound by sulfur atom in PAPS results in the cleavage of the phosphosulfate bond. Typical sulfonate conjugation reactions include the following:

(a) Oxygen atom of hydroxyl group in alcohol or phenol is conjugated with sulfonate to form a sulfonate derivative:

$$\Phi - OH + PAPS \rightarrow \Phi - O - SO_3H + PAP$$

(b) Sulfonate binds to N atom of aromatic amine to form sulfonate conjugate:

$$\Phi\text{–}NH_2 + PAPS \rightarrow \Phi\text{–}\underset{\displaystyle H}{\underset{|}{N}}\text{–}SO_3H + PAP$$

A variety of functional groups such as hydroxyl group of aliphatic and aromatic foreign compounds are substrates for STs. Many foreign compounds including aromatic and aliphatic alcohols are also involved in glucuronidation and also undergo sulfonation. Sulfonation is not limited to compounds that contain a hydroxyl group. Other compounds involving sulfonation conjugation include aromatic amines such as aniline.

The introduction of negative charge (SO_3^-) to a xenobiotic affects not only its aqueous solubility but also its interaction with transport proteins such as ATP-dependent multidrug resistance proteins. Sulfonation is an effective means of decreasing the toxic activities of foreign compounds; however, the sulfonate is an electron withdrawing group, and in certain cases sulfonate conjugates are chemically unstable, which can degrade to form potent electrophilic species. Nevertheless, in general, sulfonation conjugation is a detoxifying reaction, resulting in facilitating the excretion of foreign compounds from the body.

7.2.4 N-*Acetyltransferase*

N-Acetyltransferases (NATs) are cytosolic enzymes found in the liver and other tissues. Two functional NAT genes are present in human, NAT1 and NAT2, which are closely related proteins with an active site cysteine residue. There are over 25 allelic variants of NAT1 and NAT2. NATs catalyze the transfer of the acetyl group of acetyl-coenzyme A (acetyl-CoA) to aromatic amine or hydrazine group of a foreign compound, where acetyl-CoA acts as the cofactor of the reaction, such as acetyl group of acetyl-CoA binding to N atom of amine to form acetyl conjugate:

$$\Phi - NH_2 + \text{Acetyl-CoA} \rightarrow \Phi - NH - COCH_3 + CoA$$

In the NAT-catalyzed metabolism, aromatic amine or hydrazine group is converted to aromatic amide or hydrazide, respectively. The acetyl group of acetyl-CoA is initially transferred to a cysteine residue in NAT active site and is then transferred to the amine or hydrazine group of the substrate (foreign compound).

Unlike UGT, GST, or ST, NAT-catalyzed acetylation is not associated with an increase in aqueous solubility of xenobiotics. The addition of the acetyl group often leads to a conjugate that is less soluble in water, due to the covalent binding of the acetyl group.

In certain cases, acetylation of substrate may also be an activation process, since aromatic amines can be both activated and deactivated by NATs. Activation occurs when aromatic amines are initially hydroxylated by CYP450 to form aromatic hydroxy-amines, which are then converted to form esters. However, in general, acetyl conjugation is a detoxification reaction, resulting in facilitating the excretion of foreign compounds from the body.

7.2.5 Methyltransferase

Methyltransferases are primarily cytosolic enzymes but are also present in the endoplasmic reticulum. Methyltransferases are a class of enzymes that catalyze the transfer of methyl group of *S*-adenosyl-1-methionine to the substrates of methyltransferases. Several methyltransferases have been characterized, including catechol *O*-methyltransferase, thiopurine methyltransferase, and histamine *N*-methyltransferase.

Methylation is usually not a major metabolic route in xenobiotic metabolism. The conjugation reaction catalyzed by methyltransferase involves the transfer of the methyl group attached to the sulfonium ion of *S*-adenosylmethionine (SAM) to the functional group of the substrate (foreign compounds) to form a methyl conjugate, where SAM acts as the methyl donor and the functional group of the xenobiotic acts as the acceptor. Xenobiotics can undergo O-methylation, N-methylation, or S-methylation through *S*-adenosylmethionine-dependent conjugation reactions.

The functional groups of foreign compounds that are involved in methyl conjugation of phenol (O-methylation), aromatic amines (N-methylation), and sulfhydryl-containing compound (S-methylation) include:

(a) Oxygen atom of hydroxyl group in phenol is conjugated with methyl group of *S*-adenosylmethionine to form methyl conjugate:

$$\Phi - OH + SAM \rightarrow \Phi - O - CH_3 + SAH$$

(b) Methyl group of SAM attaches to N atom of amine to form methyl conjugate:

$$\Phi\text{–NH–R} + SAM \rightarrow \Phi\text{–N(}CH_3\text{)–R} + SAH$$

(c) Methyl group attached to the sulfonium ion in SAM binds to the S atom of thio compound to form methyl conjugate:

$$\Phi - SH + SAM \rightarrow \Phi - S - CH_3 + SAH$$

Similar to acetylation catalyzed by *N*-acetyltransferases, methylation generally decreases the water solubility of xenobiotics. Methyl conjugation is usually a detoxification reaction, which results in facilitating the excretion of foreign compounds.

Human methyltransferases have been reported to exhibit inherited variations, which may contribute to individual's variation in the metabolism of foreign compound. The methylation of inorganic arsenic to monomethyl-arsenic acid and dimethyl-arsenic acid has been considered to be the major pathway for inorganic arsenic biotransformation and detoxification.

7.2.6 *Acyltransferase*

Acyltransferase is a transferase enzyme that acts on acyl group. The enzyme present in mammalian mitochondria exists in two distinct forms. They are involved in the conjugation of acyl portion of carboxyl group with various amino acids. Acyltransferase conjugation metabolism usually occurs in the liver.

Acyltransferase-catalyzed conjugation reaction carries out the conjugation of carboxylic acid in xenobiotics with CoA in acyl-CoA to produce an acyl-CoA thioester, which then reacts with the amino group of an amino acid to form an amide linkage, such as the reaction of amino acid with a carboxylic acid group to form an amide:

$$\Phi\text{–COOH} + \underset{\displaystyle \overset{|}{R}}{\text{HN-CH}_2\text{COOH}} \rightarrow \Phi\text{–CO–}\underset{\displaystyle \overset{|}{R}}{\text{N}}\text{-CH}_2\text{COOH} + H_2O$$

where acyl-CoA serves as the cofactor for acyltransferase.

Amino acid conjugation of foreign compounds containing carboxylic acid is considered to be an alternative to glucuronidation conjugation. Acyltransferase-catalyzed amino acid conjugation of carboxylic acid generally results in detoxification, leading to facilitation of xenobiotics excretion. However, in certain cases, it may not be just so straightforward.

7.3 Non-conjugation Detoxification Enzymes

Non-conjugation detoxification enzymes are mainly quinone reductases and epoxide hydrolases. Reduction of quinones catalyzed by quinone reductase, a flavoprotein, to form hydroquinone is an important detoxification metabolism.

Epoxide hydratase-catalyzed reactions convert an epoxide residue to two hydroxyl residues through dihydroxylation reaction to form diols, such as in the case of transforming epoxide-containing lipids to 1,2-diols.

7.3.1 *Quinone Reductase*

Quinone reductase is one of detoxification enzymes that are involved in the non-conjugation metabolism of a wide variety of foreign compounds. Reduction of electrophilic quinones catalyzed by quinone reductase is an important detoxification pathway. The enzyme is accompanied by NADH or NADPH to act on a quinone. The reaction products are $NADP^+$ and hydroquinone:

$$\begin{array}{c} O \\ \| \\ \Phi \\ \| \\ O \end{array} \quad + \quad 2\,NADPH \quad \rightarrow \quad \begin{array}{c} OH \\ | \\ \Phi \\ | \\ OH \end{array} \quad + \quad 2\,NADP^+$$

Quinones are among the toxic products of oxidative metabolism of aromatic hydrocarbons. Quinone reductase displays a broad specificity for structurally diverse hydrophobic quinines, resulting in facilitating the metabolism of quinones.

7.3.2 *Epoxide Hydrolase*

Epoxide hydrolase catalyzes the hydrolysis of epoxides to the corresponding less reactive diols, making epoxide hydrolases as detoxifying enzymes. Epoxides are organic compounds that consist of a three-membered oxygen. Epoxide hydrolase is capable of inactivating a large number of structurally different, highly reactive epoxides and hence is an important part of the enzymatic defense mechanism of living organism against adverse effects of foreign compounds.

Epoxide hydrolases exist in two forms. Soluble epoxide hydrolase is expressed in the cytosol, while microsomal epoxide hydrolase is present in the endoplasmic reticulum. Epoxide hydrolases catalyze hydrolytic reactions for epoxides. As a detoxification enzyme, mammalian microsomal epoxide hydrolase exhibits a high expression level in the liver and a broad spectrum of substrates.

Oxidative reactions catalyzed by CYP450 can generate epoxides as metabolic intermediates during the metabolism of foreign compounds that contain unsaturated carbon-carbon bonds. The generated epoxide intermediates are unstable and chemically reactive and have been implicated as ultimate mutagenic and carcinogenic initiators. Epoxide hydrolases are essential in detoxifying foreign compounds that contain unsaturated carbon-carbon bonds, resulting in the inactivation of drugs and other xenobiotics.

Being important in the defense against adverse effects of foreign compounds, epoxide hydrolase is capable of inactivating structurally different, highly reactive epoxides. Hydrolase catalyzes the hydration of chemically reactive epoxides to the corresponding dihydrodiol products as shown below. The products can then be conjugated and excreted from the body:

```
      O                                      OH
     / \                                      |
Φ–C–C–R'     +     H2O      →      Φ–C–C–R'
                                        |
                                       OH
```

Bibliography

Arand M, Cronin A, Adamska M (2005) Oesch F. Epoxide hydrolases: structure, function, mechanism, and assay. Methods Enzymol 400:569-588.

Benson AM, Hunkeler MJ, Talalay P (1980) Increase of NAD(P)H:quinone reductase by dietary antioxidants: possible role in protection against carcinogenesis and toxicity. Proc Natl Acad Sci U S A 177:5216–5220

Chen CH (2012) Activation and detoxification enzymes: functions and implications. Springer, New York

Decker M, Arand M, Cronin A (2009) Mammalian epoxide hydrolases in xenobiotic metabolism and signalling. Arch Toxicol 83:297–318

Dekant W, Vamvakas S (1993) Glutathione-dependent bioactivation of xenobiotics. Xenobiotica 23:873–887

Evans D (1992) N-acetyltransferase. In: Kalow W (ed) Pharmacogenetics of drug metabolism. Pergamon, New York

Fretland AJ, Omiecinski CJ (2000) Epoxide hydrolases: biochemistry and molecular biology. Chem Biol Interact 129:41–59

Gamage N, Barnett A, Hempel N et al (2006) Human sulfotransferases and their role in chemical metabolism. Toxicol Sci 90:5–22

Grant DM, Blum M, Meyer UA (1992) Polymorphisms of N-acetyltransferase genes. Xenobiotica 9–10:1073–1081

Hayes JD, Flanagan JU, Jowsey IR (2005) Glutathione transferases. Annu Rev Pharmacol Toxicol 45:51–88

Kato R, Yamazoe Y (1994) Metabolic activation of N-hydroxylated metabolites of carcinogenic and mutagenic arylamines and arylamides by esterification. Drug Metab Rev 26:413–429

King C, Rios G, Green M, Tephly T (2000) UDP-glucuronosyltransferases. Curr Drug Metab 1:143–161

Klaassen CD, Boles JW (1997) Sulfation and sulfotransferases 5: the importance of 3′-phosphoadenosine 5′-phosphosulfate (PAPS) in the regulation of sulfation. FASEB J 11:404–418

Mannervik B (1988) Danielson UH. Glutathione transferases--structure and catalytic activity. CRC Crit Rev Biochem 23:283-337.

Meech R, Mackenzie PI (1997) Structure and function of uridine diphosphate glucuronosyl transferases. Clin Exp Pharmacol Physiol 24:907–915

Miners JO, Knights KM, Houston JB et al (2006) In vitro-in vivo correlation for drugs and other compounds eliminated by glucuronidation in humans: pitfalls and promises. Biochem Pharmacol 71:1531–1539

Negishi M, Pedersen LG, Petrotchenko E et al (2001) Structure and function of sulfotransferases. Arch Biochem Biophys 390:149–157

Rao PV, Krishna CM, Zigler JS Jr (1992) Identification and characterization of the enzymatic activity of zeta-crystallin from guinea pig lens. A novel NADPH:quinone oxidoreductase. J Biol Chem 267:96–102

Weinshilboum RM, Otterness DM, Szumlanski CL (1999) Methylation pharmacogenetics: catechol O-methyltransferase, thiopurine methyltransferase, and histamine N-methyltransferase. Annu Rev Pharmacol Toxicol 39:19–52

Wilce MC, Parker MW (1994) Structure and function of glutathione S-transferases. Biochim Biophys Acta 1205(1):1–18

Wildfang E, Zakharyan RA, Aposhian HV (1998) Enzymatic methylation of arsenic compounds. VI Characterization of hamster liver arsenite and methylarsonic acid methyltransferase activities in vitro. Toxicol Appl Pharmacol 152:366–375

Wood TC, Salavagionne OE, Mukherjee B et al (2006) Human arsenic methyltransferase (AS3MT) pharmacogenetics: gene resequencing and functional genomics studies. J Biol Chem 281:7364–7373

Talalay P (2000) Chemoprotection against cancer by induction of phase 2 enzymes. Biofactors 12:5–11

Zakharyan RA, Wildfang E, Aposhian HV (1996) Enzymatic Methylation of Arsenic Compounds. Toxicol Appl Pharmacol 140:77–84

Chapter 8
Detoxification Enzyme-Catalyzed Conjugation Reactions

Functionalization of foreign compounds catalyzed by activation enzymes results in increasing the solubility of parent compounds. The functionalized foreign compounds further undergo conjugation reactions catalyzed by detoxification enzymes. Phase II detoxification enzyme-catalyzed reactions are generally considered as detoxifying metabolism. The previous chapter describes detoxification enzymes and a variety of functional groups or atoms of foreign compounds that are targets of detoxification enzyme-catalyzed reactions, such as phenol, epoxide, polyphenol, carboxylic acid, and amino acid.

The present chapter discusses chemical reactions that are catalyzed by detoxification enzymes, including conjugation enzyme-catalyzed reactions (uridine 5-′-diphospho-glucuronosyltransferase, glutathione *S*-transferase, sulfotransferase, arylamine acetyltransferase, and methyltransferase) as well as non-conjugation enzyme-catalyzed reactions (quinone reductase and epoxide hydrolase). The functional groups involved in activation enzyme-catalyzed conjugation reactions are summarized in Table 8.1.

8.1 Detoxification Enzyme Reactions

Foreign compound detoxifying reactions are catalyzed by both conjugation enzymes and non-conjugation enzymes, which are described below.

8.1.1 Conjugation Enzymes

(a) Uridine 5′-diphospho-glucuronosyltransferase
    Foreign compounds functionalized by activation enzymes such as CYP450 can further undergo conjugation reactions catalyzed by detoxification enzymes, such

C.-H. Chen, *Xenobiotic Metabolic Enzymes: Bioactivation and Antioxidant Defense*,
https://doi.org/10.1007/978-3-030-41679-9_8

Table 8.1 Functional groups involved in conjugation reactions

Detoxification enzyme	Conjugation reaction	Functional group
UDP-glucuronosyltransferase	Glucuronidation	–OH, SH, $–NH_2$, CH_2, –COOH, NHOH
Glutathione *S*-transferase	Glutathione	Epoxide, halide, $–NO_2$
Sulfotransferases	Sulfation	–OH, $–NH_2$
N-Acetyltransferases	Acetylation	–OH, $–NH_2$, $–SO_2NH_2$
Methyltransferases	Methylation	–OH, $–NH_2$, –SH

Xenobiotic —Activation (such as CYP450)→ Functionalization —Detoxification (such as UGT)→ Conjugation —Export (MRP)→ Excretion

Fig. 8.1 Metabolism of xenobiotics: activation, detoxification, and excretion

as uridine 5′-diphospho-glucuronosyltransferase (UGT), to form glucuronide conjugates. Glucuronidation generates a variety of O-, N-, S-, and C-containing glucuronides. Glucuronidation of foreign compounds occurs in the liver, intestinal mucosa, and kidney. The resulted glucuronide conjugates are polar, facilitating their excretion from the body, as illustrated in Fig. 8.1. The excretion of glucuronide conjugates from the cells is carried out by ATP-dependent export pumps, such as multidrug resistance proteins (MRP), through urine and bile.

Glucuronidation is an important step in the metabolism of aromatic amines. UGT catalyzes the transfer of glucuronic acid (GA) from uridine 5′-diphospho-glucuronic acid (UDP-GA) to a foreign compound containing oxygen, nitrogen, sulfur, or carboxyl functional group. O-, N-, S- and C-glucuronide conjugation reactions are shown below:

$$\mathrm{R-OH + UDP-GA \rightarrow UDP + R-O-GA}$$

$$\mathrm{R-SH + UDP-GA \rightarrow UDP + R-S-GA}$$

$$\mathrm{R-NH_2 + UDP-GA \rightarrow UDP + R-NH-GA}$$

$$\mathrm{R-CH_2 + UDP-GA \rightarrow UDP + R-CH-GA}$$

$$\mathrm{R-COOH + UDP-GA \rightarrow UDP + R-COOGA}$$

$$\underset{\mathrm{OH}}{\underset{|}{\mathrm{R-NH}}} + \mathrm{UDP-GA} \rightarrow \mathrm{UDP} + \underset{\mathrm{OH}}{\underset{|}{\mathrm{R-N-GA}}}$$

where R denotes an aliphatic or aromatic group. In glucuronidation, the functional group of a substrate is hydroxyl (–OH), sulfate (–SH), amine ($-NH_2$), carbonyl ($-CH_2$), carboxylic (–COOH), or hydroxylamine (–NHOH). Glucuronidation is an important step in the detoxifying metabolism of foreign compounds containing these functionalized groups.

(b) Glutathione *S*-transferase

Glutathione *S*-transferase (GST) contributes to the metabolism of foreign compounds by catalyzing the reactions that conjugate glutathione (GSH) with functionalized xenobiotics derived from activation metabolism. GST plays an important role in the detoxification of a broad range of toxic foreign compounds, particularly those that may lead to cytotoxicity or mutagenic events, such as aflatoxin B_1 and benzo[a]pyrene. The resulting conjugates generally are less reactive and more water soluble and are ready for excretion in urine or bile.

GST isoforms catalyze the conjugation of GSH with an electrophilic substrate to form a thioester bond between the sulfur atom of GSH and the substrate. Conjugation with GSH occurs via a sulfhydryl group to the electrophilic center on a foreign compound. Foreign compounds that undergo GST-catalyzed conjugation reactions include alkyl- and aryl-halides, quinines, epoxides, isothiocyanates, unsaturated carbonyls, and nitro compounds. Detoxifying reactions involving glutathione conjugation of aliphatic halide, epoxide, and nitro-containing compounds are shown below:

```
R-C-C      +     GSH   →    R-C-C
  \ /                         | |
   O                          GSOH
```

$$R - Cl + GSH \rightarrow GS - R + H^+ + Cl^-$$

$$R - NO_2 + GSH \rightarrow GS - R + H^+ + NO_2^{\,-}$$

(c) Sulfotransferases

Sulfotransferase (ST)-catalyzed reaction conjugates the functional group of reactive intermediate or metabolite with the sulfo moiety. The most common sulfate donor is 3′-phosphoadenosine-5′-phosphosulphate (PAPS). The sulfate group is transferred from a donor molecule to an acceptor molecule such as alcohol or amine. Similar to glutathione conjugation, the addition of a sulfate group facilitates the excretion of a foreign compound from the cells mediated by MRP. Sulfate conjugates are excreted mainly in the urine.

A sulfonation reaction involves the transfer of sulfate group ($SO_3^{\,-}$) from PAPS to amine, hydroxylamine, or alcohol. Sulfonation conjugate formation is known to occur with aromatic or aliphatic amines, phenols, as well as primary, secondary, and tertiary alcohols. Sulfotransferase-catalyzed reactions involving alcohols, phenols, and aliphatic and aromatic amines are presented below:

$$R - OH + PAPS \rightarrow R - O - SO_3H + PAP$$

$$R - NH_2 + PAPS \rightarrow R - NH - SO_3H + PAP$$

where PAP denotes 3-phosphoadenosine 5-phosphate. Sulfate conjugation is the most important pathway in the metabolism of phenols.

(d) Acyltransferase

Acyltransferase catalyzes conjugation reactions that act upon the acyl portion of the carboxyl group (–COOH) in foreign compounds, leading to the formation of an amide or peptide bond between the carboxyl group and the amino group. The conjugation reaction requires an initial activation of xenobiotic to a CoA derivative. The resulting acyl-CoA subsequently reacts with an amino acid, giving rise to acylated amino acid conjugate and CoA.

Amino acid conjugation is an alternative conjugation process for carboxylic acid-containing xenobiotics. The resulting amino acid conjugates are generally excreted from the body by urinary. The conjugation of carboxylic acid-containing xenobiotics with endogenous amino acids, such as benzoic acid with glycine, phenylacetic acid with glutamine, xanthurenic acid with serine, and 4-nitrobenzoic acid with arginine, is an important pathway in the metabolism of carboxyl acid-containing foreign compounds.

Typical acyltransferase-catalyzed conjugation reactions are shown below. The reactions occur in two steps: the initial activation of the carboxyl group to yield reactive acyl-CoA thioester, followed by the transfer of acyl to the amino group in an amino acid:

Step 1: $\Phi - \text{COOH} + \text{Acyl - CoA} \rightarrow \Phi - \text{CO(CoA - Acyl)} + H_2O$

Step 2: $\Phi\text{–CO(CoA-Acyl)} + NH_2\text{CHRCOOH} \rightarrow \Phi\text{–CO(NHCHRCOOH)} + \text{Acyl-CoA}$

Overall: $\Phi - \text{COOH} + NH_2\text{CHRCOOH} \rightarrow \Phi - \text{CONHCHRCOOH} + H_2O$

where Φ denotes the aromatic portion of benzoic, phenylacetic, or xanthurenic acid and R represents the side chain portion of glycine, serine, or arginine.

(e) *N*-acetyltransferases

N-Acetyltransferase has been known to be important in the detoxification of drugs.

N-Acetyltransferase (NAT) catalyzes the transfer of an acetyl group from acetyl-CoA (AcCoA) to the terminal nitrogen of arylamine. Arylamine contains an aromatic hydrocarbon that has at least one amine group attached to it. Acetylation couples an amino group with the acetyl moiety, resulting in formation of acetylated derivatives, which are generally less water soluble than the parent compound.

Acetyltransferase-catalyzed reaction occurs in two stages. In the first stage, the enzyme is acetylated by acetyl-CoA. The second stage is the transfer of the acetyl group to the acceptor in foreign compound such as arylamine. Typical NAT-catalyzed conjugation reactions are shown in the following:

First step: $\text{AcCoA} \overset{\text{NAT}}{\rightarrow} \text{NAT} - \text{AcCoA}$

Second step: $\text{NAT} - \text{AcCoA} + \Phi - NH_2 \rightarrow \Phi - \text{NH} - \text{COCH}_3 + \text{NAT} + H_2O$

Overall: $\Phi - NH_2 + \text{AcCoA} \rightarrow \Phi - \text{NH} - \text{COCH}_3 + H_2O$

where the acceptor is arylamine or arylhydroxylamine.

(f) Methyltransferases

Methyltransferase catalyzes the reaction that transfers the methyl group from the donor to the acceptor (substrate). The reaction uses a reactive methyl group bound to sulfur in *S*-adenosyl methionine (SAM) as the methyl donor. SAM is a common enzyme cofactor. SAM-dependent methyltransferases act on a wide variety of small foreign compounds.

Hydroxyl (–OH), amino ($–NH_2$), and thiol (–SH) groups are potentially metabolized through methylation. Methylation also occurs on nucleic bases in DNA and amino acids in proteins. For proteins, the methylation reaction occurs on nitrogen atoms, either on the N terminus or side-chain position. Methyltransferase-catalyzed reactions involving the transfer of methyl group to O, N, or S-nucleophile are shown below:

$$RCH_2NH_2 + SAM \rightarrow RCH_2NH - CH_3 + SAH$$

$$\Phi - OH + SAM \rightarrow \Phi - O - CH_3 + SAH$$

$$HSCH_2CH_2OH \ + \ SAM \ \rightarrow \ CH_3 - SCH_2CH_2OH \ + \ SAH$$

where SAM and SAH denote *S*-adenosylmethionine and *S*-adenosyl-L-homocysteine, respectively.

8.1.2 Non-conjugation Enzymes

Detoxification enzymes, such as UGT, GST, ST, NAT, and methyltransferase, are traditionally referred as the enzymes that catalyze the conjugation reactions. However, the terminology has gradually become broader in scope, toward including a number of non-conjugation enzymes such as quinone reductase, epoxide hydrolase, and heme oxygenase-1. These non-conjugation enzymes are described below:

(a) Quinone reductase

Quinones are among the toxic products of oxidative metabolism of aromatic hydrocarbons. Reduction of electrophilic quinones by quinone reductase is an important detoxification pathway. Quinone reductase exhibits a broad specificity for structurally diverse hydrophobic quinones. Using NADH or NADPH as a source of reductant, quinone reductase catalyzes the reduction of quinones to phenols as shown below:

$$O = \Phi = O + NADH + H^+ \rightarrow O = \Phi - OH + NAD^+$$
$$\rightarrow HO - \Phi - OH + NAD^+$$

where O=Φ=O (*P*-benzoquinone) is reduced to O=Φ–OH (semiquinone) and then further reduced to HO–Φ–OH (hydroquinone):

$$O = \Phi = O + 2NADPH \rightarrow HO - \Phi - OH + 2NADP$$

(b) Epoxide hydrolase
Epoxides are compounds containing a three-member ring. Epoxides are ethers, but the three-membered ring gives them unusually reactivity, because of the highly polarized oxygen-carbon bonds in addition to a highly strained ring. Some reactive epoxides are responsible for electrophilic reactions with cell components such as proteins and DNA, leading to toxic and carcinogenic effects.

Epoxide hydrolases catalyze the hydrolysis reaction of the epoxide ring in alkene or arene compounds to form dihydrodiol. The addition of water to an epoxide produces 1,2-diols, where water is the co-substrate, and the reaction is energetically favorable. The hydration of benzo[a]pyrene or allylbenzene oxide forms benzo(a) pyrene diol or allylbenzene diol. A typical epoxide hydrolase-catalyzed reaction is shown below:

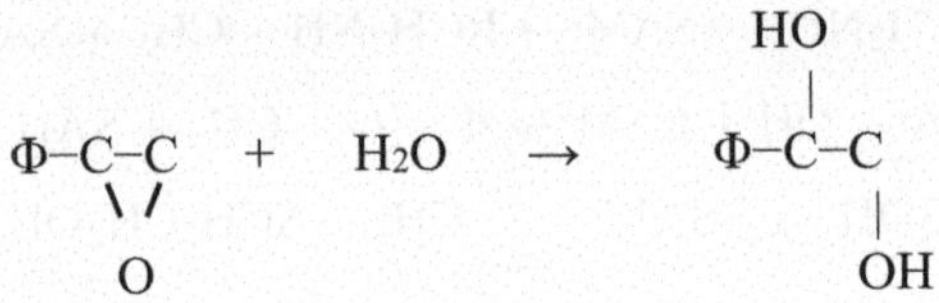

where Φ denotes an aromatic or an aliphatic group.

(c) Heme oxygenase
Heme oxygenase (HO) is the enzyme that catalyzes the degradation of heme. It cleaves heme to form biliverdin, carbon monoxide, and ferrous iron, where biliverdin can be further metabolized to bilirubin via biliverdin reductase. Two isoforms of heme oxygenase (HO-1 and HO-2) are expressed in mammalian cells:

$$\text{Heme} + O_2 \rightarrow \text{Biliverdin} + Fe^{2+} + \text{Co}$$

8.2 Conjugation Reaction at Specific Atom or Group

Detoxification enzyme-catalyzed reaction can occur at a specific atom or a functional group in foreign compound, such as O, N, C, and S atom or carboxylic group. Typical catalytic actions are presented below.

8.2.1 Conjugation at O Atom

(a) Uridine diphosphate-glucuronosyltransferase
Oxygen atom of hydroxyl group in phenol is conjugated with glucuronic acid (GA) to form a glucuronide derivative:

$$\Phi - \text{OH} + \text{UDP} - \text{GA} \rightarrow \Phi - \text{O} - \text{GA} + \text{UDP}$$

(b) Sulfotransferase
Oxygen atom of hydroxyl group in alcohol or phenol is conjugated with sulfonate to form a sulfonate derivative:

$$\Phi - OH + PAPS \rightarrow \Phi - O - SO_3H + PAP$$

(c) Methyltransferase
Oxygen atom of hydroxyl group in phenol is conjugated with methyl group of *S*-adenosylmethionine to form methyl conjugate:

$$\Phi - OH + SAM \rightarrow \Phi - O - CH_3 + SAH$$

where methyl group in SAM binds to O atom in phenol.

8.2.2 *Conjugation at N Atom*

(a) Uridine diphosphate (UDP)-glucuronosyltransferases
Glucuronic acid (GA) attaches to N atom of amine to form a glucuronide conjugate:

$$\begin{array}{c} OH \\ | \\ \Phi\text{–}N \\ | \\ H \end{array} + \text{UDP–GA} \rightarrow \begin{array}{c} OH \\ | \\ \Phi\text{–}N \\ | \\ GA \end{array} + \text{UDP}$$

(b) Sulfotransferase
Sulfonate binds to N atom of aromatic amine to form sulfonate conjugate:

$$\Phi\text{–}NH_2 + PAPS \rightarrow \begin{array}{c} \Phi\text{–}N\text{–}SO_3H \\ | \\ H \end{array} + PAP$$

(c) Methyltransferase
Methyl group of SAM attaches to N atom of amine to form methyl conjugate:

$$\Phi\text{–}NH\text{–}R + SAM \rightarrow \begin{array}{c} CH_3 \\ | \\ \Phi\text{–}N\text{–}R \end{array} + SAH$$

(d) Acetyltransferase
Acetyl group of acetyl-CoA binds to N atom of amine to form acetyl conjugate:

$$\Phi - NH_2 + \text{Acetyl-CoA} \rightarrow \Phi - NH - COCH_3 + CoA$$

where Acetyl-CoA serves as the cofactor of the enzyme.

8.2.3 *Conjugation at C Atom*

(a) Glutathione *S*-transferase
Reduced glutathione (GSH) binds to C atom of epoxide to form a glutathione conjugate:

$$\begin{array}{ccccc} \text{O} & & & & \text{OH} \\ /\backslash & & & & | \\ \Phi{-}C{-}C{-}R' & + & GSH & \rightarrow & R{-}C{-}C{-}R' \\ |\ \ | & & & & |\ \ | \\ H\ \ H & & & & H\ \ SG \end{array}$$

8.2.4 *Conjugation at S Atom*

(a) Uridine diphosphate-glucuronosyltransferases
Glucuronic acid (GA) binds to the S atom of thio compound to form a glucuronide conjugate:

$$R - SH + UDP - GA \rightarrow R - S - GA + UDP$$

(b) Methyltransferase
The methyl group attached to the sulfonium ion in SAM binds to the S atom of thio
compound to form methyl conjugate:

$$\Phi - SH + SAM \rightarrow \Phi - S - CH_3 + SAH$$

8.2.5 *Conjugation of Carboxylic Acid*

(a) Uridine diphosphate-glucuronosyltransferase
Glucuronic acid (GA) reacts with carboxylic acid to form a glucuronide conjugate:

$$\Phi - COOH + UDP - GA \rightarrow \Phi - COOGA + UDP$$

(b) *N*-acyltransferase
An amino acid reacts with a carboxylic acid group of foreign compound to form an amide, where acyl-CoA serves as the cofactor for *N*-acyltransferase:

$$\begin{array}{ccccccc} \Phi{-}COOH & + & HN{-}CH_2COOH & \rightarrow & \Phi{-}CO{-}N{-}CH_2COOH & + & H_2O \\ & & | & & | & & \\ & & R & & R & & \end{array}$$

Bibliography

Armstrong RN (1987) Enzyme-catalyzed detoxification reactions: mechanisms and stereochemistry. CRC Crit Rev Biochem 22(1):39–88

Armstrong RN (1991) Glutathione S-transferases: reaction mechanism, structure, and function. Chem Res Toxicol 4(2):131–140

Baez S, Segura-Aguilar J, Widersten M, Johansson AS, Mannervik B (1997) Glutathione transferases catalyse the detoxification of oxidized metabolites (o-quinones) of catecholamines and may serve as an antioxidant system preventing degenerative cellular processes. Biochem J. 324 (pt 1):25–28

Berhane K, Widersten M, Engström A, Kozarich JW, Mannervik BD (1994) Detoxification of base propenals and other alpha, beta-unsaturated aldehyde products of radical reactions and lipid peroxidation by human glutathione transferases. Proc Natl Acad Sci 91(4):1480–1484

Butterworth M, Lau SS, Monks TJ (1996) Estradiol metabolism by hamster hepatic microsomes. Implications for the catechol-O-methyl transferase-mediated detoxification of catechol estrogens. Drug Metab Dispos 24(5):588–594

Capel ID, Millburn P, Williams RT (1974) The conjugation of 1- and 2-naphthols and other phenols in the cat and pig. Xenobiotica 4(10):601–615

Chasseaud LF (1979) The role of glutathione and glutathione S-transferases in the metabolism of chemical carcinogens and other electrophilic agents. Adv Cancer Res 29:175–274

Chen CH (2012) Activation and detoxification enzymes: functions and implications. Springer, New York

Ciotti M, Lakshmi VM, Basu N, Davis BB, Owens IS, Zenser TV (1999) Glucuronidation of benzidine and its metabolites by cDNA-expressed human UDP glucuronosyltransferases and pH stability of glucuronides. Carcinogenesis 20(10):1963–1969

Coles B, Ketterer B (1990) The role of glutathione and glutathione transferases in chemical carcinogenesis. Crit Rev Biochem Mol Biol 25(1):47–70

Cook IT, Duniec-Dmuchowski Z, Kocarek TA, Runge-Morris M, Falany CN (2009) Hydroxycholesterol sulfation by human cytosolic sulfotransferases: formation of monosulfates and disulfates, molecular modeling, sulfatase sensitivity, and inhibition of liver x receptor activation. Drug Metab Dispos 37(10):2069–2078

Duffel MW, Marshal AD, McPhie P, Sharma V, Jakoby WB (2001) Enzymatic aspects of the phenol (aryl) sulfotransferases. Drug Metab Rev 33(3–4):369–395

Fjellstedt TA, Allen RH, Duncan BK, Jakoby WB (1973) Enzymatic conjugation of epoxides with glutathione. J Biol Chem 248(10):3702–3707

Green MD, Tephly TR (1998) Glucuronidation of amine substrates by purified and expressed UDP-glucuronosyltransferase proteins. Drug Metab Dispos 26(9):860–867

Hammock BD, Storms DH, Grant DF (1997) Epoxide hydrolases. In: Guengerich FP (ed) Biotransformation. Elsevier Sciences, New York, pp 283–305

Hayes JD, Judah DJ, McLellan LI, Neal GE (1991) Contribution of the glutathione S-transferases to the mechanisms of resistance to aflatoxin B1. Pharmacol Ther 50(3):443–472

Hirom PC, Idle JR, Millburn P, Williams RT (1977) Glutamine conjugation of phenylacetic acid in the ferret. Biochem Soc Trans 5(4):1033–1035

Hitchcock M, Smith JN (1964) Detoxification of aromatic acids in arachnids: arginine, glutamic acid and glutamine conjugations. Biochem J 93(2):392–400

Honma W, Shimada M, Sasano H, Ozawa S, Miyata M, Nagata K, Ikeda T, Yamazoe Y (2002) Phenol sulfotransferase, ST1A3, as the main enzyme catalyzing sulfation of troglitazone in human liver. Drug Metab Dispos 30(8):944–949

Hutt AJ, Caldwell J (1990) Amino acid conjugation. In: Mulder GJ (ed) Conjugation reactions in drug metabolism. An integrated approach. Taylor and Francis, London, pp 273–305

Ioannides C (2002) Enzyme systems that metabolise drugs and other xenobiotics. Wiley, Chichester

Jernström B, Seidel A, Funk M, Oesch F, Mannervik B (1992) Glutathione conjugation of trans-3,4- dihydroxy 1,2-epoxy 1,2,3,4-tetrahydrobenzo[c]phenanthrene isomers by human glutathione transferases. Carcinogenesis 13(9):1549–1555

Jin CJ, Miners JO, Burchell B, Mackenzie PI (1993) The glucuronidation of hydroxylated metabolites of benzo[a]pyrene and 2-acetylaminofluorene by cDNA-expressed human UDP-glucuronosyltransferases. Carcinogenesis 14(12):2637–2639

Keen JH, Jakoby WB (1978) Glutathione transferases. Catalysis of nucleophilic reactions of glutathione. J Biol Chem 253(16):5654–5657

Ketterer B (1986) Detoxification reactions of glutathione and glutathione transferases. Xenobiotica 16(10–11):957–973

Kiehlbauch CC, Lam YF, Ringer DP (1995) Homodimeric and heterodimeric aryl sulfotransferases catalyze the sulfuric acid esterification of N-hydroxy-2-acetylaminofluorene. J Biol Chem 270 (32):18941–18947

King CD, Rios GR, Green MD, Tephly TR (2000) UDP-glucuronosyltransferases. Curr Drug Metab 1(2):143–161

Klaassen CD, Boles JW (1997) Sulfation and sulfotransferases 5: the importance of 3-′-phosphoadenosine 5′-phosphosulfate (PAPS) in the regulation of sulfation. FASEB J 11 (6):404–418

Kolm RH, Danielson UH, Zhang Y, Talalay P, Mannervik B (1995) Isothiocyanates as substrates for human glutathione transferases: structure-activity studies. Biochem J. 311(pt 2):453–459

Lemke LE, McQueen CA (1995) Acetylation and its role in the mutagenicity of the antihypertensive agent hydralazine. Drug Metab Dispos 23(5):559–565

McGurk KA, Brierley CH, Burchell B (1998) Drug glucuronidation by human renal UDP-glucuronosyltransferases. Biochem Pharmacol 55(7):1005–1012

McLellan LI, Wolf CR, Hayes JD (1989) Human microsomal glutathione S-transferase. Its involvement in the conjugation of hexachlorobuta-1,3-diene with glutathione. Biochem J 258 (1):87–93

Minchin RF, Hanna PE, Dupret JM, Wagner CR, Rodrigues-Lima F, Butcher NJ (2007) Arylamine N- acetyltransferase I. Int J Biochem Cell Biol 39(11):1999–2005

Minchin RF (1995) Acetylation of p-aminobenzoylglutamate, a folic acid catabolite, by recombinant human arylamine N-acetyltransferase and U937 cells. Biochem J 307:1–3

Mulder GJ (ed) (1990) Conjugation reactions in drug metabolism. An integrated approach. Taylor and Francis, London

Mulder GJ, Jakoby WB (1990) Sulfation. In: Mulder GJ (ed) Conjugation reactions in drug metabolism. An integrated approach. Taylor and Francis, London, pp 107–161

Negishi M, Pedersen LG, Petrotchenko E, Shevtsov S, Gorokhov A, Kakuta Y, Pedersen LC (2001) Structure and function of sulfotransferases. Arch Biochem Biophys 390(2):149–157

Orzechowski A, Schrenk D, Bock-Hennig BS, Bock KW (1994) Glucuronidation of carcinogenic arylamines and their N-hydroxy derivatives by rat and human phenol UDP-glucuronosyltransferase of the UGT1 gene complex. Carcinogenesis 15(8):1549–1553

Parker MH, McCann DJ, Mangold JB (1994) Sulfation of di- and tricyclic phenols by rat liver aryl sulfotransferase isozymes. Arch Biochem Biophys 310(2):325–331

Ritter JK (2000) Roles of glucuronidation and UDP-glucuronosyltransferases in xenobiotic bioactivation reactions. Chem Biol Interact 129(1–2):171–193

Servin AL, Wicek D, Oryszczyn MP, Jacquot C, Lussiana JP, Christinaki H, Viel C (1987) Metabolism of 6,7-dimethoxy 4-(4′-chlorobenzyl)isoquinoline. II. Role of liver catechol O-methyltransferase and glutathione. Xenobiotica 17(12):1381–1391

Temellini A, Mogavero S, Giulianotti PC, Pietrabissa A, Mosca F, Pacifici GM (1993) Conjugation of benzoic acid with glycine in human liver and kidney: a study on the interindividual variability. Xenobiotica 23(12):1427–1433

Weber WW, Vatsis KP (1993) Individual variability in p-aminobenzoic acid N-acetylation by human N- acetyltransferase (NAT1) of peripheral blood. Pharmacogenetics 3(4):209–212

Weisiger RA, Pinkus LM, Jakoby WB (1980) Thiol S-methyltransferase: suggested role in detoxification of intestinal hydrogen sulfide. Biochem Pharmacol 29(20):2885–2887

Zeldin DC, Wei S, Falck JR, Hammock BD, Snapper JR, Capdevila JH (1995) Metabolism of epoxyeicosatrienoic acids by cytosolic epoxide hydrolase: substrate structural determinants of asymmetric catalysis. Arch Biochem Biophys 316(1):443–451

Zeldin DC, Kobayashi J, Falck JR, Winder BS, Hammock BD, Snapper JR, Capdevila JH (1993) Regio- and enantiofacial selectivity of epoxyeicosatrienoic acid hydration by cytosolic epoxide hydrolase. J Biol Chem 268(9):6402–6407

Chapter 9
Reactive Intermediates Generated from Bioactivation

Some foreign compounds are acutely or potentially toxic, while others initially exhibit no intrinsic toxicity, but become harmful after metabolic activation. Although metabolism of foreign compounds aims to produce aqueous-soluble and less toxic metabolites, thus facilitating their excretion from the body; however, in many cases, metabolic conversion produces highly chemically reactive intermediates or metabolites. Consequently, many toxic effects of foreign compounds do not result from the parent compounds, instead from reactive intermediates or metabolites that are formed inside cells during metabolic activation.

Xenobiotic toxicity associated with enzymatic conversion to reactive intermediate or metabolite is commonly referred to as bioactivation. Lipophilic foreign compounds are metabolized in two phases: activation and detoxification. A functional group such as hydroxyl or carboxyl is introduced into a foreign compound by activation enzyme-catalyzed reaction. Subsequently, a conjugation molecule such as glucuronic, sulfuric, or amino acid is added to the functionalized foreign compound by detoxification enzyme-catalyzed reaction.

Bioactivation is brought about mainly by functionalization reactions catalyzed by activation enzymes. In minor cases, it may be brought about by conjugation reactions catalyzed by detoxification enzymes. Thus, the toxic effect of a foreign compound is determined by not only the chemical nature of foreign compound but also by the enzyme reaction that metabolizes it. Reactive intermediate formation during metabolic bioactivation is an important mechanism attributing to foreign compound-mediated toxicity.

Figure 9.1 illustrates foreign compound metabolism in association with reactive intermediate formation. The figure reveals that activation metabolism either produces inactive metabolites or yields highly chemically reactive intermediates which are capable of interacting with cellular components, if they are not quickly detoxified by detoxification metabolism.

The increased toxicity of metabolic intermediate or metabolite is primarily the results of conversion of xenobiotics into chemically reactive species such as reactive oxygen species and free radicals.

C.-H. Chen, *Xenobiotic Metabolic Enzymes: Bioactivation and Antioxidant Defense*,
https://doi.org/10.1007/978-3-030-41679-9_9

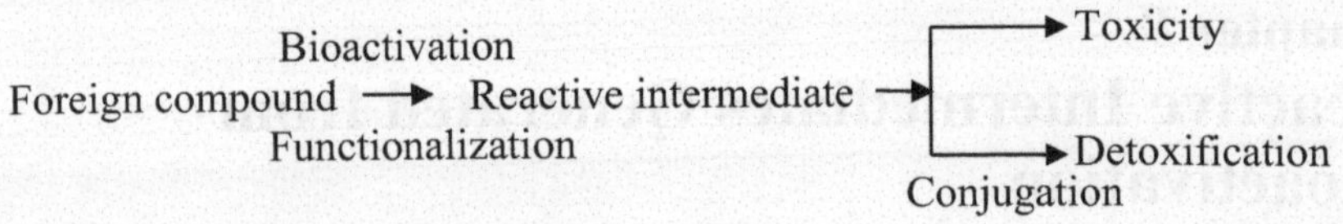

Fig. 9.1 Reactive intermediates generated by bioactivation

The generated reactive intermediates are electrophilic in nature. Electrophilic intermediates are consisted of nonionic and cationic species. Nonionic electrophilic intermediates are such as aldehydes, ketones, epoxides, quinones, sulfoxides, nitroso compounds, and acyl halides. Cationic electrophilic intermediates include carbonium ions and nitrenium ions. The metabolism of foreign compounds into nucleophiles is not common. Many of them are further activated and converted to electrophiles.

9.1 Reactive Intermediates

Many drugs and other chemicals can be activated to chemically reactive intermediates through metabolic enzymatic oxidation reactions. For example, the oxidation of a thiophene ring catalyzed by CYP450 enzyme to form reactive sulfoxide intermediates. Sulfoxide can covalently bind to thiol in proteins, affecting protein functions. Other functional groups including olefins, alkynes, alkylamines, and furans can also be oxidized to generate reactive intermediates.

Although functionalization reactions catalyzed by activation enzymes convert many xenobiotics into unharmful substances, however, a significant number of functionalized foreign compounds are substantially more reactive than the parent compounds. While foreign compound metabolism serves as an important defense biological process, in some cases, it leads to producing adverse effects within the organism.

Damage of cells and tissues can occur if chemically reactive intermediates are produced in sufficient quantities and are not promptly stabilized by defense mechanism. It is therefore important to minimize the presence of reactive intermediates in intracellular levels and to quickly remove them from the body.

Most chemical toxicants or carcinogens require metabolic bioactivation catalyzed by activation enzymes to exert their toxic effects. However, the consequence of chemically reactive intermediate is dependent on not only the individual reaction but also the complex pathway and the cellular state. Consequently, not all foreign compounds that form reactive intermediates are toxic. Many of them are not.

9.1.1 *Reactive Oxygen/Nitrogen Species*

Reactive intermediates generated from bioactivation of foreign compounds can interact with oxygen molecules to produce reactive oxygen species. Figure 9.2 illustrates the generation of reactive oxygen species from reactive intermediates.

Bioactivation / Oxygen

Foreign compound → Reactive intermediate → Reactive oxygen species

Fig. 9.2 Formation of reactive oxygen species

Reactive oxygen species, including highly reactive cationic electrophiles such as carbonium, iminium, and nitrenium ions, can therefore be considered as sorts of metabolic reactive intermediates.

Reactive oxygen species are continuously produced during metabolic processes of foreign compounds and are meanwhile detoxified by a variety of defense mechanisms. When the process of generation is more favorable than that of detoxification, damages of cellular macromolecules occur. Reactive oxygen species are also implicated in cell signaling and in the immune system as a way to attack pathogens.

Reactive nitrogen species include nitric oxide (NO) and its by-products such as nitrate (NO_3^-), nitrite (NO_2^-), peroxynitrites (^-OONO), and peroxynitrite ($ONOO^-$). Ionic reactive nitrogen species are such as nitroxyl anion (NO^-) and nitrosyl cation (NO^+). Nitric oxide can react with O_2^- to give peroxynitrite, a stronger oxidant than O_2^- or NO, which can oxidize cellular components (proteins, lipids, and DNA).

9.1.2 *Free Radicals*

Free radicals are also formed during metabolic bioactivation of foreign compounds and are therefore also considered as sorts of reactive intermediates. Superoxide anion radical ($O_2^{-\bullet}$) and hydroxyl radical ($HO^\bullet$) are typical free radicals. Similar to reactive oxygen species, free radicals are able to interact with cellular components (proteins, lipids, and DNA), resulting in damages to cellular molecules and their functions. Peroxidase is a major enzyme that catalyzes the formation of free radicals. The mechanism underlying free radical formation from xenobiotic reactive intermediates is different from that generated from aerobic electron transfer chains in living organisms.

9.1.3 *Positively or Neutral Charged Electrophiles*

Many foreign compounds undergoing oxidation reactions catalyzed by CYP450 enzymes produce electrophilic reactive intermediates. Meanwhile, many other products of xenobiotic oxidation reactions are nonreactive metabolites. Reactive intermediates produced by metabolic activation are often electrophilic species. Electrophilic species are compounds that are attracted to electrons and therefore are Lewis acid. An electrophile carries a partial positive charge or possess an atom not having an octet of electrons. Electrophilic species include cations such as H^+ and NO^+, as well as polarized neutral molecules such as alkyl halides, acyl halides, and carbonyl compounds. Electrophiles can take part in chemical reactions that involve binding to nucleophiles in cellular components.

9.2 Bioactivation of Foreign Compounds

9.2.1 *Drugs*

Metabolic bioactivation to form reactive intermediates has been believed to be the major contributing mechanism to drug-induced toxicity. Drug-induced toxicity is a leading cause of drug withdrawal from clinical use. It also represents a considerable impediment to the development of new drugs. Table 9.1 presents typical drugs that form reactive intermediates through metabolic bioactivation.

A potential source of drug-associated toxicity is largely resulted from bioactivation catalyzed by CYP450 enzymes, such as in the case of nimesulide, a nitroaromatic sulfoanilide anti-inflammatory drug. Nimesulide is bioactivated by CYP2C to form quinone reactive intermediate, which is associated with nimesulide-mediated liver injury. Other drugs that are associated with hepatotoxicity contain antipsychotics chlorpromazine and thioridazine, which undergo CPY450-mediated bioactivation to form quinone imine intermediates.

Drugs that contain a nitroaromatic moiety, such as tolcapone, nilutamide, flutamide, nitrofurantoin, have also been found to associate with liver injury. Nitroreductive bioactivation that produces nitro anion radical, nitroso intermediate, and N-hydroxy derivative.

Other reactive intermediates derived from metabolic oxidation include trovafloxacin, a fluoroquinolone antibiotic drug, which can form covalent adducts to hepatic proteins, resulting in damage to liver tissue.

Tamoxifen, a breast cancer drug, was approved as a chemopreventive agent for women at high risk for breast cancer. However, several studies have reported that the use of tamoxifen leads to increased risk of endometrial cancer in humans. Tamoxifen metabolism involves hydroxylation to yield 4-hydroxytamoxifen, which may be further oxidized to form a quinone methide.

Moreover, carboxylic acid-containing drugs are metabolized to form thioester derivatives that can covalently bind to tissue proteins, leading to potential toxicity.

Thiophene-containing anticancer drug can be oxidized by CPY450s to sulfoxide, which can covalently bind to thiol nucleophiles. The formation of thioketo aldehyde as electrophilic intermediate in metabolism of suprofen (a nonsteroidal anti-inflammatory agent) catalyzed by CYP450 enzyme was obtained by characterizing pyridazine adduct.

Table 9.1 Typical drugs that form toxic reactive intermediates

Typical drugs	Classification	Reactive intermediate
Tamoxifen	Breast cancer drug	Quinone methide
Desipramine	Antidepressants	N-Dealkylated compounds
Nimesulide	Anti-inflammation	Diiminoquinone
Chlorpromazine; thioridazine	Antipsychotic	Quinone imine
Trovafloxacin	Antibiotic	Aldehyde
Suprofen	Anti-inflammation	Thioketo-aldehyde
Carbamazepine	Idiosyncratic drug	9-Acridine carboxaldehyde

Carbamazepine, a widely used anticonvulsant, is associated with a wide range of adverse reactions including alteration of immune function. Carbamazepine is oxidized by activated neutrophils to several metabolites, leading to covalent binding of the drug to the cells. The reactive metabolite responsible for this binding is 9-acridine carboxaldehyde.

9.2.2 *Other Chemicals*

Metabolism of anisidine forms diimine metabolite which is subsequently hydrolyzed to form a quinone imine. Both diimine and quinone imine metabolites are reactive toward nucleophiles. Arylamine derived from industries and cigarette smoke undergoes bioactivation catalyzed by activation enzymes to form reactive intermediate, diazonium ion. Butadiene is metabolized by CYP450-dependent monooxygenases to form monoepoxide and diepoxide intermediates.

Humans are exposed to capsaicinoid through the consumption of peppers. Metabolism of capsaicinoid catalyzed by CPY450 enzyme generates quinone methide and quinone intermediates. Dopamine undergoes monoamine oxidase-catalyzed oxidative deamination to generate 3,4-dihydroxyphenyl-acetaldehyde, a neurotoxin that has been implicated in the pathogenesis of Parkinson's disease. Bioactivation of furan involves reactive intermediate, *cis*-2-butene-1,4-dial, which is a hepatotoxicant.

Biotransformation of hydroquinone forms benzoquinone and quinone-thioether metabolites, which have impact on nephrotoxicity and hepatotoxicity. CYP450-mediated oxidation of pulegone generates menthofuran, which is further oxidized to form electrophilic reactive intermediates, menthofuran epoxide and γ-ketoenal. Both of these intermediates form adducts to hepatocellular proteins.

3-Methylindolea is a lung toxin found in cigarette smoke. Its metabolic bioactivation generates reactive intermediates, hydroxyl, methyl, and methylene indolenines. The oxidation of 2-naphthylamine catalyzed by prostaglandin H synthase forms reactive electrophile, 2-imino-1-naphthoquinone, which can covalently bind to DNA. Sudan I [1-(phenylazo)-2-naphthol] is an industrial dye. CYP450 and peroxidases are capable of oxidizing Sudan I to form reactive intermediate (benzenediazonium cation), which was found to covalently modify DNA.

Table 9.2 summarizes typical chemicals that form toxic reactive intermediates during metabolic activation.

9.2.3 *Herbs and Dietary Supplements*

There is an increasing interest in applying herbal medicines and dietary supplements for preventing or treating various chronic diseases. Plant dietary supplements are natural origin, which are assumed to be safer than prescription drugs. While most

Table 9.2 Typical chemicals that form toxic reactive intermediates

Typical chemicals	Classification	Reactive intermediate
Anisidine	Carcinogen	Diimine and quinone imines
Arylamine	Toxin	Diazonium ion
Butadiene	Carcinogen	Epoxybutane
Capsaicinoid	Toxin	Quinone methide and quinone
Dopamine	Neurotransmitter	Aldehyde
3,4-Dihydroxyphenylacetic acid	Neurotoxin	Aldehyde
Furan	Hepatotoxin	*cis*-2-Butene-1,4-dial
Hydroquinone	Toxin	Quinone-thioether
Menthofuran	Hepatotoxin	Gamma-ketoenal
3-Methylindole	Pneumotoxin	*2,3-Epoxy-3-methylindolenine*
2-Naphthylamine	Carcinogen	2-Imino-1-naphthoquinone
Pulegone	Hepatotoxin	Menthofuran, menthofuran epoxide
Sudan I industrial dye	Carcinogen	Hydroxylated metabolites

Table 9.3 Typical chemicals of herb and dietary supplement chemicals that form toxic reactive intermediate

Typical herb or dietary supplement	Reactive intermediate
Aristolochic acid	Nitrenium ion
Diterpenoid	Epoxide

plant dietary supplements can be considered being safe, a few of them contain chemical compounds that can be converted to reactive intermediates, such as quinones, quinone methides, and epoxides, leading to toxicity.

Table 9.3 presents typical chemicals of herb and dietary supplement that form reactive intermediates.

Accumulated data suggest a role of reactive intermediates in herbal toxicity. There are increasing clinical reports on the organ toxicities associated with consumption of herbal medicines. The nitro group of aristolochic acid undergoes metabolism catalyzed by CYP450 or peroxidases to form reactive cyclic nitrenium ion. The furan ring of diterpenoid is oxidized by CYP3A4 to generate reactive epoxide. Some naturally occurring flavonoids undergo metabolic activation through hydroxylation and sulfation, resulting in generating reactive intermediates that are capable of forming DNA adducts.

Reactive intermediates following herbal bioactivation was hypothesized to covalently bind to cellular proteins and DNA, leading to toxicity. Moreover, reactive intermediates generated from bioactivation of some herbs are also capable of irreversibly inhibiting CYP450 enzymes, leading to CYP450 inactivation, which may explain some reported herb-drug interactions.

9.3 Metabolic Enzyme-Mediated Reactive Intermediate Formation

CYP450 is the most important activation enzyme with respect to the generation of biological reactive intermediates, including epoxides, carbonium, iminium, and nitrenium ions and free radicals. Other major activation enzymes such as peroxidase and flavin monooxygenases also participate in catalytic functionalization reactions that produce reactive intermediates. In minor cases, detoxification enzyme-catalyzed conjugation reactions may be involved in reactive intermediate formation. Non-conjugation enzyme may also be involved, such as epoxide hydrolase catalyzes the formation of benzo[a]pyrene 7,8-dihydrodiol-9,10-epoxide, a benzo[a]pyrene intermediate.

9.3.1 *Functionalization Reactions*

Functionalization metabolism of drugs and other chemicals convert parent compounds to form hydrophilic species. Oxidative metabolism, catalyzed by activation enzymes mainly cytochrome P450 and peroxidase, plays an important role in bioactivation. Table 9.4 presents typical reactive intermediates generated by functionalization reactions.

Table 9.4 Functionalization reaction-mediated reactive intermediates

Foreign compound	Activation enzyme	Reactive intermediate
Drugs		
Acetaminophen	UGT	*N*-Acetyl-*P*-benzoquinone imine
Menadione	CYP1A2	Semiquinone radical
Phenylbutazone	PHS	Carbon-centered/peroxyl radical
Toxins		
Aflatoxin B_1	CYP3A4/peroxidase	Aflatoxin B_1-8,9-epoxide
4-Ipomeanol	CYP4B1	α,β-Unsaturated dialdehyde
Other chemicals		
2-Acetylaminofluorene	CYP2A13	*N*-Hydroxy-acetylaminofluorene
Bromobenzene	CYP2A6	Bromobenzene 3,4 oxide
Benzo[a]pyrene (BaP)	CYP2A13/epoxide hydrolase	BaP 7,8-dihydrodiol-9,10-epoxide
2-Naphthylamine	CYP2S1	*N*-hydroxy-naphthylamine
N-Nitrosodimethylamine	CYP2E1	Carbonium ion
Polycyclic aromatics (PAH)	Epoxide hydrolase	PAH diol-epoxides

Table 9.5 Conjugation reaction-mediated reactive intermediates

Foreign compound	Conjugation enzyme	Reactive intermediate
2-Aminonaphthalene	UGT	*N*-Hydroxy-2-naphthylamine
Benzidine	NAT	Acetyloxyarylamine nitrenium ion
Dibromoethane	GST	Episulfonium ion
N-Hydroxy-2-aminofluorene	NAT	Nitrenium ion
7-Hydromethyl,12-methyl-benz[a] anthracene	SULT	Carbonium ion

9.3.2 Conjugation Reactions

In certain cases, conjugation reactions catalyzed by detoxification enzymes are also involved in the generation of reactive intermediates. Table 9.5 presents some typical conjugation reaction-mediated reactive intermediates, which are catalyzed by UDP-glucuronosyltransferase (UGT), sulfotransferase (SULT), glutathione *S*-transferase (GST), or *N*-acetyltransferase (NAT).

9.4 Oxidation Reaction-Mediated Reactive Intermediates

The main route of metabolic activation of foreign compounds to form electrophilic reactive intermediates is the oxidation reactions catalyzed by activation enzymes, such as cytochrome P450, peroxidases, flavin-monooxygenases, amine oxidases, and xanthine oxidase.

Among these enzymes, CYP450 family plays the most significant role. CYP450 uses molecular oxygen to catalyze a variety of oxidative reactions, including hydrocarbon hydroxylation, olefin epoxidation, heteroatom oxygenation, dealkylation of ethers, thioethers, and amines.

Another important enzyme, peroxidase, catalyzes the oxidation of a wide variety of organic and inorganic substrates. CYP450 and peroxidase break down many foreign compounds into electrophiles that are capable of impairing vital cellular functions by covalently binding to proteins and DNA as well as inducing lipid peroxidation. The functional groups of foreign compounds that can be oxidized to reactive intermediates include alkene (olefin), alkyne (acetylene), alkylamine, furan, thiophene, *p*-aminophenol, and phenothiazine. These CYP450 and peroxidase-catalyzed oxidation reactions are briefly described below.

9.4.1 *Cytochrome P450-Catalyzed Oxidation Reactions*

(a) Reaction occurring at oxygen atom

CYP450 family is the most important enzyme involved in a wide range of oxidation reactions in the metabolism of foreign compounds. CYP450 enzymes are hemoproteins that catalyze a range of oxidative C–H functionalization reactions including aliphatic and aromatic hydroxylation. The most common reaction catalyzed by CYP450 is a monooxygenase reaction, which inserts one atom of oxygen into the aliphatic position of an organic substrate to yield phenols, ketones, or epoxides. CYP450 enzymes also catalyzes the oxidation of alcohols to carbonyl compounds and sulfides to sulfoxides:

$$R'RCHOH \rightarrow RR'CO + H_2O$$

$$RCH_2SR \rightarrow RCH_2SOR$$

$$ROCH_3 \rightarrow ROH + HCHO$$

where R denotes aliphatic hydrocarbon and may also denote aromatic hydrocarbon.

CYP450-dependent monooxygenase system can also convert unsaturated aliphatic and aromatic hydrocarbons to epoxides during oxidative metabolism. Olefins (alkenes) are converted to epoxides by oxygen insertion, where an oxygen atom is inserted into a C=C bond of unsaturated aliphatic or aromatic hydrocarbons to form an epoxide:

$$\text{R-C=C-R'} \quad + \quad O_2 \quad \rightarrow \quad \text{R-}\overset{\text{O}}{\overset{/\backslash}{\text{C-C}}}\text{-R'}$$

where the side group R or R′ represents aliphatic hydrocarbon and may also denote aromatic hydrocarbon.

(b) Reaction occurring at nitrogen atom

Enzymatic oxidation reactions occurring at nitrogen atom include N-hydroxylation and N-oxidation. N-Hydroxylation involves the substitution of one hydrogen atom in an amino group with a hydroxy group, while N-oxidation involves the addition of oxygen to an amino group:

$$2\Phi - NH - CO - CH_3 + O_2 \rightarrow 2\Phi - NOH - CO - CH_3$$

(amine is oxidized to form hydroxyl amine)

$$2\Phi - CH = NH + O_2 \rightarrow 2\Phi - CH = NOH$$

(Imine is oxidized to form oxime in N-hydroxylation)

9.4.2 Peroxidase-Catalyzed Oxidation Reactions

Besides the major oxidative enzyme CYP450, peroxidases are among other oxidative enzymes which catalyze the oxidation of a wide variety of organic and inorganic substrates. Peroxidases are heme proteins which catalyze the oxidation of substrates through the reaction with hydrogen peroxide. Hydrogen peroxidase is an enzyme found in aerobic cells, which functions are to break down hydrogen peroxide.

Horseradish peroxidase has been shown to catalyze the oxidation of 3,4-dimethoxybenzyl alcohol and benzyl alcohol to the respective aldehydes. The oxidation reaction results in hydrogen abstraction from benzylic carbon of the substrate alcohol, leading to the aldehyde product as shown below:

$$\Phi - CH_2OH + H_2O_2 \rightarrow \Phi CHO + 2H_2O$$

Peroxidase-catalyzed oxidation reactions of tricyclic antidepressants, such as desipramine and clomipramine, result in the formation of reactive intermediates (N-dealkylated compounds). Peroxidase also catalyzes oxidative metabolism of other chemicals. Bioactivation of Sudan 1 (1-(phenylazo)-2-naphthol), an industrial dye found in foods, brings about the formation of hydroxylated metabolites, leading to toxic effects by binding to DNA.

9.5 Factors Affecting Reactive Intermediate Accumulation

Metabolic reactive intermediate or metabolite that is responsible for cellular effects of the parent compound depends on not only the chemical nature of the parent compound but also the metabolic enzyme that catalyzes its activation reaction. Figure 9.3 summaries the factors that affect reactive intermediate accumulation within the cells. The figure reveals the importance of relative rates of functionalization and conjugation reactions in xenobiotics metabolism.

While the metabolism of foreign compounds is supposed to represent a protective biological process, however, if the rate of functionalization reaction is higher than that of conjugation reaction, an accumulation of metabolic reactive intermediate within the cells occurs. An imbalance between functionalization and conjugation processes often determines whether a reactive intermediate exerts its effect or not.

Metabolic reactive intermediate responsible for cellular effects of the parent compound is also affected by the genetic makeup of metabolic activation and detoxification enzymes. Genetic polymorphisms of metabolic enzymes can cause significant variations in the expression of their enzymatic activities, resulting in affecting on the formation and accumulation of reactive intermediate. Moreover, environmental factors including chemical exposure, nutrition, as well as lifestyle such as cigarette smoking and alcohol drinking are also able to modulate the activities of metabolic enzymes.

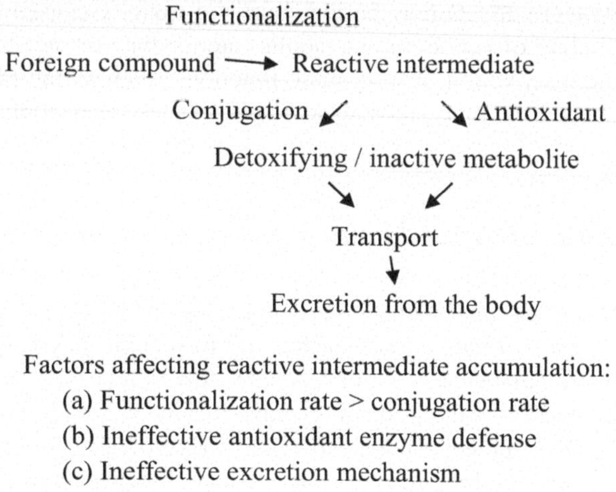

Fig. 9.3 Factors that affect reactive intermediate accumulation

In summary, the factors that affect reactive intermediate accumulation include not only the chemical nature of foreign compounds and the relative activities of metabolic enzymes but also the antioxidant defense capacity and the environment and lifestyle.

9.6 Interactions with Proteins, DNA, and Lipids

When metabolic intermediates are active and are not effectively detoxified by detoxification enzymes and antioxidants, covalently binding to proteins, lipids or DNA can occur, leading to the formation of protein and DNA adducts and the peroxidation of lipids. The interactions of chemically active intermediates with the cellular components can be either specific or non-specific. Specific interaction involves the binding of reactive intermediate to a specific site in a receptor, enzyme, or transport protein.

The liver is not only the principal site for the activation of foreign compounds but also the center of production of reactive intermediates and the primary site of toxic effects. However, reactive intermediates may also be exported to other tissues and cause toxic effects there.

9.6.1 Protein Adducts

The primary binding sites of protein adducts are sulfur-containing amino acid residues (cysteine and methionine), nitrogen of lysine, or histidine residues. Although not all metabolic intermediate bindings to protein are toxic in nature, protein adducts may be the source of toxicity caused by foreign compounds. The binding of xenobiotic metabolic intermediate or metabolite to protein can result in the inactivation of enzymatic functions, such as the binding of microcystins, produced from blue-green algae, to phosphatase leads to the inactivation of the enzyme.

9.6.2 DNA Adducts

Metabolic intermediate, particularly reactive oxygen species, binding to DNA has drawn more attention than binding to protein. DNA binding occurs when reactive intermediate reacts with DNA base. Any of four bases in DNA may participate in the interaction. Among them, guanine residue is the predominant base for the binding.

DNA adduct is not always stable. Hydrolysis can cleave the bond between purine and electrophilic reactive intermediate. Electron-dense sites, like nitrogen and oxygen atoms in nucleic acid, are primary binding sites. DNA adduct formation can affect DNA base pairing. When DNA undergoes replication, erroneous base pairing can result in base mismatch. If base mismatch is not repaired during DNA replication, mutation can occur.

9.6.3 Lipid Peroxidation

Lipid peroxidation refers to the oxidative degradation of membrane lipids. Cell damages occur when free radicals react with membrane lipids. Lipid peroxidation most often affects polyunsaturated fatty acids that contain multiple double bonds between methylene groups. The mechanism of lipid peroxidation consists of three major steps: initiation, propagation, and termination. Initially, fatty acid free radical is produced when reactive oxygen species such as OH· reacts with unsaturated lipid. The resulting unstable fatty acid free radical can react with oxygen to yield peroxyl-fatty acid radical, which can then react with another fatty acid free radical to produce a different fatty acid radical and lipid peroxide. Such propagation process proceeds by a chain reaction mechanism.

Bibliography

Amacher DE (2006) Reactive intermediates and the pathogenesis of adverse drug reactions: the toxicology perspective. Curr Drug Metab 7:219–229

Anders MW (2007) Chemical toxicology of reactive intermediates formed by the glutathione-dependent bioactivation of halogen-containing compounds. Chem Res Toxicol 21:145–159

Anders MW (1985) Bioactivation of foreign compounds. Academic, New York

Anderson D (2001) Genetic and reproductive toxicity of butadiene and isoprene. Chem Biol Interact 135–136:65–80

Baird WM, Hooven LA, Mahadevan B (2005) Carcinogenic polycyclic aromatic hydrocarbon-DNA adducts and mechanism of action. Environ Mol Mutagen 45:106–114

Boelsterli UA (2007) Mechanistic toxicology. CRC Press, Boca Raton

Boelsterli UA, Ho HK, Zhou S, Leow KY (2006) Bioactivation and hepatotoxicity of nitroaromatic drugs. Curr Drug Metab 7(7):715–727

Bogdanffy MS, Taylor ML (1993) Kinetics of nasal carboxylesterase-mediated metabolism of vinyl acetate. Drug Metab Dispos 21:1107–1111

Bolton JL (2002) Quinoids, quinoid radicals, and phenoxyl radicals formed from estrogens and antiestrogens. Toxicology 177(1):55–65

Boyd JA, Eling TE (1987) Prostaglandin H synthase-catalyzed metabolism and DNA binding of 2-naphthylamine. Cancer Res 47(15):4007–4014

Chandrasekara A, Shahidi F (2011) Inhibitory activities of soluble and bound millet seed phenolics on free radicals and reactive oxygen species. J Agric Food Chem 59:428–436

Chen CH (2012) Activation and detoxification enzymes: functions and implications. Springer, New York

Dekant W, Vamvakas S (1993) Glutathione-dependent bioactivation of xenobiotics. Xenobiotica 23:873–887

Dietz BM, Bolton JL (2011) Biological reactive intermediates (BRIs) formed from botanical dietary supplements. Chem Biol Interact 192:72–80

Eaton DL, Gallagher EP (1994) Mechanisms of aflatoxin carcinogenesis. Annu Rev Pharmacol Toxicol 34:135–172

Furst SM, Uetrecht JP (1995) The effect of carbamazepine and its reactive metabolite, 9-acridine carboxaldehyde, on immune cell function in vitro. Int J Immunopharmacol 17(5):445–452

Glatt H (2000) Sulfotransferases in the bioactivation of xenobiotics. Chem Biol Interact 129:141–170

Guengerich FP (2001) Common and uncommon cytochrome P450 reactions related to metabolism and chemical toxicity. Chem Res Toxicol 14:611–650

Hinson JA, Forkert PG (1995) Phase II enzymes and bioactivation. Can J Physiol Pharmacol 73:1407–1413

James LP, Capparelli EV, Simpson PM et al (2008) Acetaminophen-associated hepatic injury: evaluation of acetaminophen protein adducts in children and adolescents with acetaminophen overdose. Clin Pharmacol Ther 84:684–690

Jinsmaa Y, Florang VR, Rees JN, Mexas LM, Eckert LL, Allen EM, Anderson DG, Doorn JA (2011) Dopamine-derived biological reactive intermediates and protein modifications: implications for Parkinson's disease. Chem Biol Interact 192(1-2):118–121

Johnson WW (2008) Many drugs and phytochemicals can be activated to biological reactive intermediates. Curr Drug Metab 9(4):344–351

Kalgutkar AS, Dalvie DK, O'Donnell JP et al (2002) On the diversity of oxidative bioactivation reactions on nitrogen-containing xenobiotics. Curr Drug Metab 3:379–424

Khojasteh SC, Hartley DP, Ford KA, Uppal H, Oishi S, Nelson SD (2012) Characterization of rat liver proteins adducted by reactive metabolites of menthofuran. Chem Res Toxicol 25 (11):2301–2309

Kim SY, Suzuki N, Laxmi YR et al (2004) Genotoxic mechanism of tamoxifen in developing endometrial cancer. Drug Metab Rev 36:199–218

Khojasteh SC, Hartley DP, Ford KA, Uppal H, Oishi S, Nelson SD (2012) Characterization of rat liver proteins adducted by reactive metabolites of menthofuran. Chem Res Toxicol 25 (11):2301–2309

Koob M, Dekant W (1991) Bioactivation of xenobiotics by formation of toxic glutathione conjugates. Chem Biol Interact 77:107–136

Korobkova EA, Nemeth J, Cadougan M, Venkatratnam A, Bassit M, Azar N (2012) Reactive metabolites of desipramine and clomipramine: the kinetics of formation and reactivity with DNA. Bioorg Med Chem 20(1):340–345

Levi PE, Hodgson E (2008) Reactive metabolites and toxicity. In: Smart RC, Hodgson E (eds) Molecular and biochemical toxicology. Wiley, New York

Mao J, Dai W, Zhang S, Sun L, Wang H, Gao Y, Wang J, Zhang F (2019) Quinone-thioether metabolites of hydroquinone play a dual role in promoting a vicious cycle of ROS generation: in vitro and in silico insights. Arch Toxicol 93(5):1297–1309

McLemore TL, Litterst CL, Coudert BP et al (1990) Metabolic activation of 4-ipomeanol in human lung, primary pulmonary carcinomas, and established human pulmonary carcinoma cell lines. J Natl Cancer Inst 82:1420–1426

Medower C, Wen L, Johnson WW (2008) Cytochrome P450 oxidation of the thiophene-containing anticancer drug 3-[(quinolin-4-ylmethyl)-amino]-thiophene-2-carboxylic acid (4-trifluoromethoxy-phenyl)-amide to an electrophilic intermediate. Chem Res Toxicol 21(8):1570–1577

O'Donnell JP, Dalvie DK, Kalgutkar AS, Obach RS (2003) Mechanism-based inactivation of human recombinant P450 2C9 by the nonsteroidal anti-inflammatory drug suprofen. Drug Metab Dispos 31(11):1369–1377

Perlow RA, Kolbanovskii A, Hingerty BE et al (2002) DNA adducts from a tumorigenic metabolite of benzo[a]pyrene block human RNA polymerase II elongation in a sequence- and stereochemistry-dependent manner. J Mol Biol 321:29–47

Phillips MB, Sullivan MM, Villalta PW, Peterson LA (2014) Covalent modification of cytochrome c by reactive metabolites of furan. Chem Res Toxicol 27(1):129–135

Raucy JL, Kraner JC, Lasker JM (1993) Bioactivation of halogenated hydrocarbons by cytochrome P4502E1. Crit Rev Toxicol 23:1–20

Reilly CA, Henion F, Bugni TS, Ethirajan M, Stockmann C, Pramanik KC, Srivastava SK, Yost GS (2013) Reactive intermediates produced from metabolism of the vanilloid ring of capsaicinoids by P450 enzymes. Chem Res Toxicol 26(1):55–66

Ritter JK (2000) Roles of glucuronidation and UDP-glucuronosyltransferases in xenobiotic bioactivation reactions. Chem Biol Interact 129:171–193

Shimada T (2006) Xenobiotic-metabolizing enzymes involved in activation and detoxification of carcinogenic polycyclic aromatic hydrocarbons. Drug Metab Pharmacokinet 21:257–276

Smart RC, Hodgson E (2008) Molecular and biochemical toxicology. Wiley, New York

Smith BJ, Curtis JF, Eling TE (1991) Bioactivation of xenobiotics by prostaglandin H synthase. Chem Biol Interact 79:245–264

Sridar C, D'Agostino J, Hollenberg PF (2012) Bioactivation of the cancer chemopreventive agent tamoxifen to quinone methides by cytochrome P4502B6 and identification of the modified residue on the apoprotein. Drug Metab Dispos 40(12):2280–2288

Stadtman ER (2006) Protein oxidation and aging. Free Radic Res 40:1250–1258

Stiborova M, Schmeiser HH, Frei E, Hodek P, Martinek V (2014) Enzymes oxidizing the azo dye 1-phenylazo-2-naphthol (Sudan I) and their contribution to its genotoxicity and carcinogenicity. Curr Drug Metab 15(8):829–840

Thompson DC, Eling TE (1991) Reactive intermediates formed during the peroxidative oxidation of anisidine isomers. Chem Res Toxicol 4(4):474–481

Turrens JF (2003) Mitochondrial formation of reactive oxygen species. J Physiol 552:335–344

Weems JM, Lamb JG, D'Agostino J, Ding X, Yost GS (2010) Potent mutagenicity of 3-methylindole requires pulmonary cytochrome P450-mediated bioactivation: a comparison to the prototype cigarette smoke mutagens B(a)P and NNK. Chem Res Toxicol 23(11):1682–1690

Zhou S, Koh HL, Gao Y, Gong ZY, Lee EJ (2004) Herbal bioactivation: the good, the bad and the ugly. Life Sci 74(8):935–968

Zhou SF, Xue CC, Yu XQ, Wang G (2007) Metabolic activation of herbal and dietary constituents and its clinical and toxicological implications: an update. Curr Drug Metab 8(6):526–553

Chapter 10
Electrophilic Nature of Metabolic Reactive Intermediates

Many toxic effects of foreign compounds do not result from the parent compounds, instead from reactive intermediates or metabolites that are formed inside cells during metabolic activation. Such reactive intermediate formation during metabolic activation is an important mechanism attributing to foreign compound-mediated toxicity. While foreign compound metabolism serves as an important protective biological process, in some cases it leads to producing adverse effects within the organism.

Damage of cells and tissues can occur if chemically reactive intermediates are produced in sufficient quantities and are not promptly stabilized by defense mechanisms. In order to exert their toxic and neoplastic effects, most chemical toxicants and carcinogens require metabolic activation catalyzed by activation enzymes such as CYP450 enzymes. For instance, quinones generated from CYP450/peroxidase-catalyzed oxidation of catechol estrogens have been implicated as the ultimate carcinogens. Meanwhile, many foreign compounds are also oxidized by CYP450 enzymes to non-toxic metabolites. Thus, the same enzyme that detoxifies a foreign compound can be responsible for the activation of another one.

Reactive intermediates including reactive oxygen/nitrogen species and free radicals are able to react with neutrophilic donors from target cellular molecules, such as proteins, lipids, and DNA. Lipid peroxidation most often affects polyunsaturated fatty acids due to multiple double bonds in between. An increase in polyunsaturated fatty acid may stimulate the oxidation of potential toxicants.

Drug-induced toxicity is a leading cause of drug withdrawal from clinical use.

Chemically reactive intermediates are hence a major concern in the pharmaceutical industry. Although the mechanisms underlying drug-induced toxicities are varied, the metabolic bioactivation to form reactive intermediates or metabolites is considered as the major contributing factor. Reactive species can covalently modify important biological molecules and thereby increase the risk of adverse drug reactions or toxicity.

Screening compounds for their propensity to form reactive intermediates or metabolites as well as their reactions with nucleophilic species becomes an important part of drug development programs. The issue of chemically reactive intermediates

C.-H. Chen, *Xenobiotic Metabolic Enzymes: Bioactivation and Antioxidant Defense*,
https://doi.org/10.1007/978-3-030-41679-9_10

or metabolites in drug metabolism is a growing concern in the pharmaceutical industry due to their role in mediating drug-induced toxicities.

The activation of antioxidant response element (ARE)-nuclear factor E2-related factor 2 (Nrf2) pathway plays an important role in the defense mechanism against reactive intermediate-mediated oxidative damages to proteins, lipids, and DNA. Chemical inducers of Nrf2 are able to react with Keap1 cysteine residues, leading to the activation of ARE.

10.1 Reactive Intermediates

Reactive intermediates produced by functionalization reactions are generally electrophilic in nature. The conversion of foreign compounds into nucleophiles is relatively not common. Although some nucleophiles may be reactive, many of them are further activated by conversion to electrophiles. Electrophilic reactive intermediates preferentially form covalent bonds with nucleophiles of comparable polarizability of cellular components.

Electrophile covalently binding to nucleophilic sites results in irreversible alterations in cellular functions. Irreversible electrophilic adduction of cellular macromolecules is involved in enzyme inactivation and pathogenic processes underlying many disease conditions. Reactive oxygen/nitrogen species are generally electron-deficient species such as positively charged carbonium ion ($R–CH_2^+$) and atomic or molecular species with unpaired electrons.

The nucleophilic sites most often targeted by electrophilic reactive intermediates are the side chains on protein amino acids such as Cys, His, and Lys or aromatic nitrogen sites on DNA bases (guanine and adenine). The sulfhydryl thiol (R–SH) side chain of Cys residues is a weak nucleophile, which can be ionized to become a more reactive nucleophilic thiolate (RS^-).

The functional groups of electrophilic reactive intermediates binding to the nucleophilic sites of cellular components have been a focus of investigation. Irreversible electrophile-nucleophile interactions are prominently involved in oxidative stress and other underlying pathophysiological processes that mediate drug and chemical toxicity as well as various disease conditions.

10.2 Electrophiles

An electrophile is an electron acceptor, a cation, or a neutral compound but can never be an anion. Electrophilic intermediates of foreign compounds are capable of covalent bonding with nucleophilic sites of cellular molecules to form stable adducts with target proteins, DNA, or lipids. Covalent modifications of cellular components can initiate processes that lead to irreversibly alter cellular functions.

Most electrophiles are positively charged, have an atom carrying a partial positive charge, or possess an atom not having an octet of electrons. Typical electrophiles include cations (such as H^+ and NO^+), polarized neutral molecules (such as HCl, alkyl halides, acyl halides, and carbonyl compounds), and polarizable neutral molecules (such as Cl_2 and Br_2).

10.3 Reactive Oxygen Species

Reactive intermediates can interact with oxygen molecules to produce reactive oxygen species, which include highly reactive cationic electrophiles, such as carbonium, iminium, and nitrenium ions as well as free radicals. Peroxidases are a major enzyme that catalyzes the formation of free radicals. Free radicals formed in metabolic activation include superoxide anion radical ($O_2^{-\bullet}$) and hydroxyl radical ($HO^{\bullet}$). It should be pointed out that the mechanism underlying free radical formation from reactive intermediates of foreign compounds is different from that generated from aerobic electron transfer chains in living organisms.

Reactive intermediates and free radicals are able to interact with cellular components, resulting in damages to cellular molecules and their functions. Reactive oxygen species are continuously produced during the metabolic processes of foreign compounds. Meanwhile, they are detoxified by a variety of defense mechanisms including detoxification enzymes and antioxidants. Damages of cellular macromolecules occur when the process of generation of reactive intermediates is more favorable than that of detoxification.

10.4 Functional Groups of Electrophiles

In an electrophile consisting of a carbon with sp^2 electronic configuration (such as aldehyde and ketone), the carbon tends to contain a polarized C=X bond, where X can be oxygen, nitrogen, or sulfur. For carbonyl compounds where X is oxygen, the polarity of the carbon-oxygen bond exhibits a partial positive charge δ^+ on the carbon atom and a partial negative charge δ^- on the oxygen atom. As a result, the electron-deficient carbon atom is an electrophile, where a nucleophilic attack may occur.

Since electrophiles are electron-deficient species, an electrophile (EP) is attracted to electrons and therefore participates in a chemical reaction by bonding to a nucleophile (NP) through either positive-negative charge interaction or accepting a pair of electrons. Typical electrophile-nucleophile reactions are shown below:

Table 10.1 Functional groups of electrophiles and respective nucleophiles

Electrophile	Nucleophile
Epoxide	Thiols
Aldehyde	Amine
Ketone/aldehyde	Hydroxyl amine
Ester	Amine
Isothiocyanate	Amine

(a) Positive-negative charge interaction:

$$EP^{\delta+} + NP^{\delta-} \rightarrow EP\text{-}NP$$

(b) Interaction with accepting a pair of electrons

$$EP+ : NP \rightarrow EP : NP$$

Table 10.1 presents some functional groups of electrophiles that bind to respective nucleophiles.

10.5 Cellular Nucleophilic Sites Involving Reactive Intermediate Binding

A nucleophile donates electrons and therefore is a Lewis base. Being electron donors, nucleophiles are negatively charged, have π bonds, or possess lone pairs of electrons. Negatively charged anions are such as Br^-, OH^-, CN^-, and $RCOO^-$. Compounds that possess π bonds are such as RC=CR′ and RC=CR′. Atoms that contain lone pairs of electrons are such as $:NH_3$ and RR′C=O:, where R and R′ denote an aliphatic or aromatic group.

Nucleophilic functional groups are electron-rich atoms. They are able of donating a pair of electrons to form a new covalent bond with electrophiles. Important nucleophilic functional groups in biological molecules include hydroxyl, methyl, carbonyl, carboxyl, amino, phosphate, and sulfhydryl. These functional groups play critical roles in the formation of cellular molecules like proteins, DNA, and membrane lipids. Their involvements in electrophile-nucleophile interactions have been a subject of research focus.

10.5.1 Proteins

Chemically reactive intermediates of xenobiotics are generated from activation enzyme-catalyzed functionalization reactions, mainly during oxidative metabolism. These reactive species are electrophilic in nature and are capable of forming stable

Table 10.2 Main nucleophilic sites of cellular components

Cellular components	Nucleophilic sites
Protein	Cysteine, sulfhydryl, lysine, thiolate, and histidine amino residues
DNA	N2- and N7-guanine; N6-adenine
Lipids	Hydroxyl, carboxyl, and (−) phosphate

adducts with target proteins. Electrophilic reactive intermediates of foreign compounds can covalently bind to cellular proteins, resulting in enzyme inactivation or cell damages. These covalent protein modifications can initiate processes that lead to acute tissue injury or chronic disease.

Covalent binding of reactive intermediates or metabolites to proteins is considered to be one of the important mechanisms by which drugs can cause tissue damage. Hydroquinone is a known environmental toxicant. Its quinone-thioether metabolites, formed via the intermediate generation of 1,4-benzoquinone, elicit their toxic response through the covalent modification of target proteins and the generation of reactive oxygen species.

Table 10.2 presents major nucleophilic sites of proteins potential for electrophilic bindings with cellular components. The nucleophilic sites of proteins that are most often targeted by electrophilic intermediates are the side chains on protein amino acids, including cysteine (Cys) sulfhydryl (–SH), thiolate^{-} (–RS^{-}), lysine (Lys), and histidine (His) amino acid residues.

Sulfhydryl is a functional group consisting of a sulfur bonded to a hydrogen atom. Sulfhydryl plays a critical role in forming disulfide bonds that connect amino acids together for functional purpose in proteins structures. Thiol is the compound that contains sulfhydryl functional group. Sulfhydryl thiol (R–SH) side chain of cysteine residue is a weak nucleophile that can be ionized to become a more reactive nucleophilic thiolate (RS^{-}), which is critically involved in many biochemical processes in the cell.

Comparable polarizability has been proposed for electrophile binding to nucleophile. Strong electrophiles which can react with nucleophilic side chains of proteins include aldehydes and ketones. Electrophiles such as epoxides, quinones, quinone-imines, and quinone methides that are generated from oxidative bioactivation of phenyl, phenolic, and amino substituents are trapped to nucleophiles such as cysteine, while electrophiles, like iminiums and aldehydes generated from oxidative bioactivation of amines and alcohols, are capable of binding to nucleophiles such as cyanide and amines.

Protein covalently binding by an electrophilic reactive intermediate can initiate a variety of toxic responses. Most of these toxic responses share the ability to modify nucleophilic sulfhydryl groups. Electrophilic diverse chemicals, such as acrylamide, 2,4-dithiobiuret, and methylmercury, cause synaptic dysfunction. Synaptic activity is regulated by the redox state of cysteine-sulfhydryl groups on proteins. Electrophilic neurotoxicants may produce synaptic toxicity by modifying thiols. Highly nucleophilic cysteine-thiolate groups within the catalytic groups of synaptic proteins can act as specific targets for electrophilic neurotoxicants.

Irreversible covalent modification on proteins by electrophilic reactive intermediates can have substantial pathogenic consequences. For examples, dopamine undergoes monoamine oxidase-catalyzed oxidative deamination to form 3,4-dihydroxyphenylacetaldehyde (DOPAL). Electrophilic DOPAL reacting with protein nucleophiles may serve as a chemical trigger in pathogenesis. Aldehyde intermediate is a neurotoxicant. It has been implicated as a factor in the pathogenesis of Parkinson's disease.

Moreover, some drugs that contain a thiophene ring are oxidized by CYP450 to form reactive intermediates such as sulfoxides. Sulfoxides can covalently bond to thiol nucleophiles in proteins. Moreover, quinone-thioether metabolites of hydroquinone form 1,4-benzoquinone intermediate, which can elicit toxic response through covalent modification of target protein in the binding motifs within lysine-rich regions.

10.5.2 DNA

Numerous foreign compounds including therapeutics agents are substrates for bioactivation to produce electrophilic reactive intermediates that may also covalently modify DNA. The bases of nucleic acid components possess π electrons that are susceptible to be attacked by electrophilic reactive intermediates. Many chemical carcinogens or their reactive metabolites are electrophiles that exert their effects by covalent bonding to nucleophilic sites in DNA, resulting in the formation of altered nucleobases known as DNA adducts.

DNA adduct formation is believed to signal the onset of chemically induced carcinogenesis, which could cause toxic immune response or cancer. Such modification may lead to changes in the genetic composition of the cell, including mutation. The nucleophilic centers most often targeted by electrophilic reactive intermediates are aromatic nitrogen sites on DNA bases such as guanine.

Guanine N7 is one of the most favorable sites of DNA adduct formation. Epoxy metabolites of 1,3-butadiene are electrophilic, which can bind to nucleophilic sites in DNA to form guanine N7 adducts of epoxy butene and butanediol. Styrene oxide, a reactive metabolite of styrene, modifies DNA at nucleophilic sites on 7-guanine and 1-adenine to form styrene oxide-DNA adducts. DNA alkylation or adduct formation occurring at nucleophilic sites N7-guanine is considered as a potential biomarker for metabolic activation of carcinogens.

Moreover, quinone methides form DNA adducts, where the alkylation occurs primarily at N2- and N7-guanine and N6-adenine. Covalent adduct formation of aldehyde electrophile was reported to occur at N6-adenine of DNA bases. Major nucleophilic sites of DNA potential for electrophilic bindings are also presented in Table 10.2.

Electrophilic epoxides are produced by enzyme-catalyzed oxidation of parent compounds. Epoxides can react with DNA to form DNA adducts. Quinone metabolites are essentially formed through oxidation reactions catalyzed by CYP450 or peroxidases. Highly reactive organic electrophilic quinone species, including quinone-imines, quinone-methides, and imine-methides, have the potential to covalently attack DNA bases and thus initiate mutagenic and carcinogenic processes. Many adverse drug reactions also produce electrophilic reactive metabolites that can conjugate to nucleophilic sites of DNA.

10.5.3 Lipids

Cellular lipids include fatty acid, glyceride, sterol, and phospholipid. Fatty acid is a carboxylic acid that consists of a long aliphatic chain with a terminal carboxyl group. Glyceride is an ester formed from glycerol and fatty acid, where glycerol has three hydroxyl functional groups that can react with fatty acids. Sterol is a subgroup of steroid with a hydroxyl group and cholesterol. Phospholipids, the main component of cell membranes, are composed of glycerol, fatty acid, and phosphate group, where the phosphate group consists of a negatively charged polar head.

Among cellular lipid functional groups are hydroxyl, carboxyl, and negatively charged phosphate groups which are nucleophilic sites that can be targeted by electrophilic reactive intermediate. Table 10.2 also presents major nucleophilic sites of membrane lipids which are potential for bindings by electrophilic reactive intermediates of foreign compounds.

Double bonds in polyunsaturated fatty acids are one of the major targets of reactive oxygen species. Membrane phospholipids and triglycerides are primary targets for hydroxyl radical-mediated attack. Lipid peroxidation of acyl chains is the result of hydroxyl radical attack of fatty acyl chains of phospholipids and triglycerides.

Lipid peroxidation products represent a major group of chemically diverse compounds formed by oxidative reactions. Oxidative cleavage leads to the formation of small aliphatic and lipid-bound aldehydes and ketones. α,β polyunsaturated aldehydes are potent electrophiles that can covalently modify not only lipids but also proteins and DNA. Binding of an electrophilic intermediate to a lipid nucleophilic site can lead to cytotoxicity, resulting in altering cellular lipid functions.

10.6 Defenses Against Metabolic Reactive Intermediates

The protection against the harmful effects of electrophilic reactive intermediates or metabolites is to minimize their presence in intracellular levels. To achieve this goal, the body develops defense systems that include detoxification enzymes, antioxidant enzymes, glutathione, and vitamins E and C and carotene. These defense systems against active intermediates are discussed below:

10.6.1 Detoxification Enzymes

In order to reduce the intracellular levels of electrophilic reactive intermediates or metabolites, the body depends heavily on metabolic conjugation reactions catalyzed by detoxification enzymes, including uridine diphosphate-glucuronosyltransferases, glutathione *S*-transferases and sulfotransferases. Conjugation reactions not only detoxify reactive intermediates but also greatly increase aqueous solubility of foreign compounds (with exceptions such as methylation), thus facilitating the excretion of foreign compounds from the body.

Metabolic conjugation usually represents a detoxification process in a large majority of studies. Nevertheless, in some cases, such as the conjugation of a foreign compound with glucuronic acid, glutathione, or sulfonate, detoxification enzyme-catalyzed reactions was originally thought to represent exclusively a detoxifying process for drugs and environmental chemicals. Instead, it has been found that the resulting conjugates may not be so straightforward.

10.6.2 Antioxidant Enzymes

The body also utilizes antioxidant enzymes as the primary line of defense against free radicals produced in aerobic cellular metabolism. Among antioxidant enzymes present in the living organisms includes superoxide dismutase, which converts the superoxide radical to form hydrogen peroxide and oxygen, while catalase and glutathione peroxidase work simultaneously with glutathione to reduce hydrogen peroxide to water and oxygen.

In addition to serving as a primary line of defense against free radicals, antioxidant enzymes can also effectively break down reactive oxygen species derived from metabolic conversion of foreign compounds. For examples, glutathione peroxidase catalyzes the reaction in which glutathione breaks down hydrogen peroxide and organic peroxide, while superoxide dismutase catalyzes the reactions that break down superoxide anions.

10.6.3 Glutathione

Electrophilic reactive intermediates or metabolites are molecules that contain an intramolecular center of low electron density (positive center), which prefers to interact with nucleophilic species that contains intramolecular center of high electron density (negative center). Besides detoxification enzymes and antioxidant enzymes, the body also relies on endogenous molecules to remove electrophilic intermediates. Glutathione, a nucleophilic species, is the most important endogenous molecules for neutralizing free radicals produced by foreign compound metabolism.

Glutathione reacts with chemically active intermediates or metabolites to produce water-soluble compounds, thus facilitating their excretion from the body through either urine or bile. When glutathione acts as antioxidant, it is oxidized to glutathione disulfide. The function of glutathione as an antioxidant does not involve detoxification enzyme. Unlike in glutathione *S*-transferase-catalyzed conjugation reaction, glutathione acts as a cofactor.

10.6.4 Vitamins E and C and Carotene

The body also utilizes small antioxidant molecules as the primary line of defense in detoxifying free radicals produced in aerobic cellular metabolism. Small antioxidant molecules such as vitamins E and C and carotene can neutralize free radicals by accepting or donating an electron, leading to the elimination of the state of an unpaired electron in free radicals. Vitamin E protects cell membranes from oxidation damage by free radicals. Vitamin C scavenges free radicals and works with vitamin E to quench free radicals. Beta-carotene is an effective quencher of singlet oxygen.

Bibliography

Baillie TA (2006) Future of toxicology-metabolic activation and drug design: challenges and opportunities in chemical toxicology. Chem Res Toxicol 19(7):889–893

Bolton JL, Shen L (1996) p-Quinone methides are the major decomposition products of catechol estrogen o-quinones. Carcinogenesis 17(5):925–929

Boysen G, Pachkowski BF, Nakamura J, Swenberg JA (2009) The formation and biological significance of N7-guanine adducts. Mutat Res 678(2):76–94

Chen CH (2012) Activation and detoxification enzymes: functions and implications. Springer, New York

Coon MJ, Ding XX, Pernecky SJ, Vaz AD (1992) Cytochrome P450: progress and predictions. FASEB J 6(2):669–673

Fisher AA, Labenski MT, Monks TJ, Lau SS (2011) Utilization of LC-MS/MS analyses to identify site- specific chemical protein adducts in vitro. Methods Mol Biol 691:317–326

Furst SM, Sukhai P, McClelland RA, Uetrecht JP (1995) Covalent binding of carbamazepine oxidative metabolites to neutrophils. Drug Metab Dispos 23(5):590–594

Guengerich FP (2003) Cytochrome P450 oxidations in the generation of reactive electrophiles: epoxidation and related reactions. Arch Biochem Biophys 409:59–71

Hauck AK, Bernlohr DA (2016) Oxidative stress and lipotoxicity. J Lipid Res 57(11):1976–1986

Hemeryck LY, Vanhaecke L (2016) Diet-related DNA adduct formation in relation to carcinogenesis. Nutr Rev 74(8):475–489

Hughes TB, Swamidass SJ (2017) Deep learning to predict the formation of quinone species in drug metabolism. Chem Res Toxicol 30(2):642–656

Jurva U, Holmén A, Grönberg G, Masimirembwa C, Weidolf L (2008) Electrochemical generation of electrophilic drug metabolites: characterization of amodiaquine quinoneimine and cysteinyl conjugates by MS, IR, and NMR. Chem Res Toxicol 21(4):928–935

Kalgutkar AS (2017) Liabilities associated with the formation of "hard" electrophiles in reactive metabolite. Chem Res Toxicol 30(1):220–238

Koivisto P, Peltonen K (2001) N7-guanine adducts of the epoxy metabolites of 1,3-butadiene in mice lung. Chem Biol Interact 135–136:363–372

Korobkova EA, Nemeth J, Cadougan M, Venkatratnam A, Bassit M, Azar N (2012) Reactive metabolites of desipramine and clomipramine: the kinetics of formation and reactivity with DNA. Bioorg Med Chem 20(1):340–345

Koskinen M, Vodicka P, Vodickova L, Hemminki K (2001) (32)P-postlabelling/HPLC analysis of various styrene-induced DNA adducts in mice. Biomakers 6(3):175–189

Lewis MA, Yoerg DG, Bolton JL, Thompson JA (1996) Alkylation of 2′-deoxynucleosides and DNA by quinone methides derived from 2,6-di-tert-butyl-4-methylphenol. Chem Res Toxicol 9 (8):1368–1374

LoPachin RM, Gavin T (2016) Reactions of electrophiles with nucleophilic thiolate sites: relevance to pathophysiological mechanisms and remediation. Free Radic Res 50(2):195–205

Sono M, Roach MP, Coulter ED, Dawson JH (1996) Heme-containing oxygenases. Chem Rev 96 (7):2841–2888

McEldoon JP, Dordick JS (1991) Thiol and Mn(2+)-mediated oxidation of veratryl alcohol by horseradish peroxidase. J Biol Chem 266(22):14288–14293

Meesters RJ, Duisken M, Hollender J (2007) Study on the cytochrome P450-mediated oxidative metabolism of the terpene alcohol linalool: indication of biological epoxidation. Xenobiotica 37 (6):604–617

Meunier B, de Visser SP, Shaik S (2004) Mechanism of oxidation reactions catalyzed by cytochrome P450 enzymes mechanism of oxidation reactions catalyzed by cytochrome P450 enzymes. Chem Rev 104(9):3947–3980

Nelson SD, Mitchell JR, Timbrell JA, Snodgrass WR, Corcoran GB 3rd (1976) Isoniazid and iproniazid: activation of metabolites to toxic intermediates in man and rat. Science 193:901–903

O'Donnell JP, Dalvie DK, Kalgutkar AS, Obach RS (2003) Mechanism-based inactivation of human recombinant P450 2C9 by the nonsteroidal anti-inflammatory drug suprofen. Drug Metab Dispos 31(11):1369–1377

Ortiz de Montellano PR (2010) Hydrocarbon hydroxylation by cytochrome P450 enzymes. Chem Rev 110(2):932–948

Price NE, Johnson KM, Wang J, Fekry MI, Wang Y, Gates KS (2014) Interstrand DNA-DNA cross-link formation between adenine residues and abasic sites in duplex DNA. J Am Chem Soc 136(9):3483–3490

LoPachin RM, Gavin T (2016) Reactions of electrophiles with nucleophilic thiolate sites: relevance to pathophysiological mechanisms and remediation. Free Radic Res 50(2):195–205

Skordos KW, Laycock JD, Yost GS (1998) Thioether adducts of a new imine reactive intermediate of the pneumotoxin 3-methylindole. Chem Res Toxicol 11(11):1326–1331

Sono M, Roach MP, Coulter ED, Dawson JH (1996) Heme-containing oxygenases. Chem Rev 96:2841–2887

Stiborova M, Schmeiser HH, Frei E, Hodek P, Martinek V (2014) Enzymes oxidizing the azo dye 1-phenylazo-2-naphthol (Sudan I) and their contribution to its genotoxicity and carcinogenicity. Curr Drug Metab 15(8):828–840

Sun Q, Zhu R, Foss FW Jr, Macdonald TL (2008) In vitro metabolism of a model cyclopropylamine to reactive intermediate: insights into trovafloxacin-induced hepatotoxicity. Chem Res Toxicol 21(3):711–719

Thompson CD, Kinter MT, Macdonald TL (1996) Synthesis and in vitro reactivity of 3-carbamoyl-2-phenylpropionaldehyde and 2-phenylpropenal: putative reactive metabolites of felbamate. Chem Res Toxicol 9(8):1225–1229

Thompson DC, Eling TE (1991) Reactive intermediates formed during the peroxidative oxidation of anisidine isomers. Chem Res Toxicol 4(4):474–481

Wen B, Zhou M (2009) Metabolic activation of the phenothiazine antipsychotics chlorpromazine and thioridazine to electrophilic iminoquinone species in human liver microsomes and recombinant P450s. Chem Biol Interact 181(2):220–226

Yan Z, Zhong HM, Maher N, Torres R, Leo GC, Caldwell GW, Huebert N (2005) Bioactivation of 4- methylphenol (p-cresol) via cytochrome P450-mediated aromatic oxidation in human liver microsomes. Drug Metab Dispos 33(12):1867–1876

Chapter 11
Oxidative Stress Mediated by Reactive Intermediates

Metabolic enzyme-catalyzed reactions for foreign compounds occur in two processes: activation and detoxification. Activation enzyme-mediated metabolism involves a serial of oxidation, reduction, and hydrolytic reactions, which catalyze the conversion of foreign compounds to hydrophilic species. In many cases, activation reactions come along with the generation of reactive chemical intermediates and metabolites, such as reactive oxygen species and free radicals. Detoxification enzyme-catalyzed conjugation reactions alleviate the parent compounds into inactive or less toxic and more water-soluble metabolites, hence facilitating their excretion from the body.

Reactive intermediates including reactive oxygen species and free radicals generated from activation metabolism are associated with inflammation and carcinogenesis. Uncontrolled inflammatory response can result in cell and tissue damages. Oxidative stress occurs as the production of reactive oxygen species, free radicals, or other reactive intermediates is not balanced by the capability of body's defense systems to detoxify them.

Oxidative stress can cause a significant mediation of cellular components, including proteins, DNA, and membrane lipids, leading to implicating the pathogenesis and pathological conditions. Oxidative stress-mediated cellular damages play a crucial role in the pathogenesis processes that can attribute to cardiovascular disorder, neurodegeneration, pulmonary injury, and cancers. An elevated level of chemically reactive intermediates including reactive oxygen species and free radicals requires an overwhelming response from detoxification enzymes and endogenous antioxidants such as glutathione, superoxide dismutase, and catalase.

A wide range of detoxification and antioxidant genes are regulated by the transcription factor, erythroid 2-related factor 2 (Nrf2). The activation of the Nrf2 pathway promotes and induces detoxification enzymes and antioxidant proteins to counter oxidative stress. Reactive oxygen species or chemical intermediates trigger a stage of regulatory factors, leading to the activation of the transcription factor Nrf2 in response to toxicants and carcinogens. The activation of the Nrf2 pathway has been thought to be a promising strategy for cytoprotection and chemoprevention.

C.-H. Chen, *Xenobiotic Metabolic Enzymes: Bioactivation and Antioxidant Defense*,
https://doi.org/10.1007/978-3-030-41679-9_11

11.1 Exogenous and Endogenous Reactive Intermediates

Reactive intermediates generated from metabolism of foreign compounds are exogenously related, such as the products of drugs and other chemicals. Reactive intermediates can also be endogenously related, such as those produced during mitochondrial oxidative metabolism by organisms. Discussions below address both exogenously and endogenously related reactive intermediates.

11.1.1 Exogenously Related Reactive Intermediates

In many cases, a variety of reactive chemical intermediates are formed during the course of bioactivation. Many potentially harmful foreign compounds are metabolized to chemically reactive transient species before they exert toxic actions on cellular components. Reactive chemical intermediates can covalently bind to cellular components (proteins, DNA, and membrane lipids), causing protein adduct formation, DNA transformation and mutation, and lipid peroxidation.

CYP450 enzymes play a prominent role in bioactivation metabolism of a large number of foreign compounds, including many drugs and other chemicals. For instance, CYP450 family member, CYP2E1, is expressed not only in the liver but also in other tissues including the kidney, lung, brain, and gastrointestinal tract. CYP2E1 mediates metabolism of foreign compounds, such as ethanol, acetaminophen, nitrosamines, and azo compounds, to generate reactive chemical intermediates and reactive oxygen species. Oxidative stress derived from increased CYP2E1 protein levels and induced enzymatic activity is a main cause of various liver disease conditions, such as in the case of chronic consumption of alcohol.

11.1.1.1 Drugs

Reactive thiophene intermediates generated from the oxidation or epoxidation reaction are potential cause for their adverse side effects. As a result, thiophene-containing drugs such as tienilic acid and suprofen have been withdrawn from prescription. Trovafloxacin, an antibiotic drug, is associated with many cases of hepatotoxicity. Trovafloxacin-induced hepatotoxicity may be mediated through the oxidation of cyclopropylamine to reactive intermediates, which form covalent adducts to hepatic proteins.

Aflatoxin B_1 catalyzed by CYP3A4 enzyme forms a carcinogenic intermediate, aflatoxin-8,9-epoxide, which is detoxified by the conjugation reaction catalyzed by glutathione S-transferase (GST) enzyme. However, a relatively low expression of GST for aflatoxin-8,9-epoxide is highly susceptible to toxic and carcinogenic effects of aflatoxin B_1.

Bioactivation of acetaminophen (a pain and fever relief drug) in the liver to N-acetyl-p-benzoquinone imine is known to bind to liver proteins, resulting in hepatotoxicity. Moreover, isoniazid (an antituberculosis drug) and iproniazid (an antidepressant) are oxidized by CYP450 enzymes in liver microsomes to highly reactive acylating and alkylating agents. Covalent binding of these reactive intermediates to hepatic proteins was reported to associate with hepatic cellular necrosis.

11.1.1.2 Other Chemicals

Toxic reactive intermediates also occur during metabolism of environmental chemicals. The lung is the primary target site for the toxicity of environmental pollutants including ethylene oxide, butadiene, styrene, and acrylonitrile. Although the lung has a capability of transforming such foreign compounds to reduce their potential toxicity, however, biotransformation mediated by CYP450 activation sometimes may result in the generation of more reactive, toxic reactive intermediates. Butadiene is metabolized to reactive metabolites, 1,2-epoxy-3-butene and diepoxybutane, which play a critical role in butadiene carcinogenicity.

CYP450-mediated oxidation of menthofuran, a naturally occurring hepatotoxin, is metabolically activated to reactive chemical intermediates, which are capable of covalent binding to cellular proteins. Pulegone is also a naturally occurring hepatotoxin, which can form an unsaturated gamma-ketoaldehyde as its ultimate reactive metabolite. Quinone reductase-catalyzed reduction of quinones is considered to play a protective role against quinone-mediated toxicity. However, complication may be produced during metabolism of quinones by quinone reductase.

The chemical, phthalates, commonly found in personal care products, may be an even more important route of exposure than are plastics. Phthalates may generate reactive oxygen species that affect various physiological and reproductive functions, such as accommodation male fertility by inducing a state of oxidative stress in the testes. It has been reported that both CYP450-mediated activation reaction and glutathione conjugation reaction are involved in the formation of episulfonium ion intermediate. Moreover, CYP450 and epoxide hydrolase enzymes catalyze the conversion of environmental carcinogen polycyclic aromatic hydrocarbons (PAHs) to PAH-diols and PAH diol-epoxides, which can be further converted to reactive radical cations catalyzed by CYP450 and peroxidases.

Reactive intermediates o-quinones generated from the oxidation of catechol have a variety of biological effects, such as the cytotoxic effects of phenolic botanical extracts. Poison ivy contains oily urushiol, which is oxidized to o-quinones, leading to skin toxicity through oxidative stress. Botanical o-quinones can also be formed by O-dealkylation of O-alkoxy groups, resulting in catechols being further oxidized to o-quinones.

Dopamine undergoes monoamine oxidase-catalyzed oxidative deamination to 3,4-dihydroxyphenylacetaldehyde, which is metabolized to 3,4-dihydroxyphenylacetic acid (DOPAC) via aldehyde dehydrogenase (ALDH). The aldehyde intermediate was implicated as a factor in the pathogenesis of Parkinson's disease.

Although conjugation reactions of reactive intermediates are considered as a mechanism of detoxification, however, in some cases, the formation of conjugates such as those with glutathione may lead to bioactivation, because glutathione conjugates could be further metabolized into reactive species that may cause toxic effects.

11.1.2 Endogenously Related Reactive Intermediates

Unlike exogenous metabolites that are derived from metabolism of foreign compounds like drugs, environmental chemicals, foods, and toxins, endogenously related reactive intermediates or metabolites are produced by organisms, such as metabolic products of amino acids, vitamins, and fatty acids. Compared to exogenous metabolites, oxidative stress mediated by endogenous metabolites has been less investigated, and the role of endogenous metabolites in various chronic diseases has not been adequately explored.

A typical example of endogenous metabolite-mediated chronic disease involves the metabolism of retinoids (retinal and retinol) and retinoic acid catalyzed by CYP450 enzymes, resulting in generation of epoxides as reactive intermediates. If such endogenous metabolites are not rapidly hydrolyzed by epoxide hydrolases, retinoid epoxides may act as toxicants to cause potential damage to the neurons in the macula. Genetic polymorphisms of CYP450 enzymes responsible for epoxide formation as well as epoxide hydrolases responsible for epoxides hydrolysis may serve to explain the involvement of genetic component in the process of age-related macular degeneration.

11.2 Reactive Oxygen Species (ROS)

Reactive oxygen species are reactive intermediates of molecular oxygen, which are able to oxidize DNA and interfere with the mechanisms of DNA repair, and also capable of reacting with proteins and lipids. Generation of reactive oxygen species occurs in biotransformation of products known to human daily life. Reactive oxygen species have been reported to associate with a number of tissue/organ injuries induced by metabolic activation of xenobiotics.

Metabolism of menadione (a nutritional supplement with vitamin K activity) by quinone reductase can lead to the production of OH radical in the presence of various ferric catalysts.

Extensive research has been carried out to investigate the underlying effects of reactive oxygen species-mediated oxidative stress and their contribution to the aging process. An imbalance between the generation of reactive oxygen species (ROS) and antioxidant defense systems represents the primary cause of endothelial dysfunction, leading to vascular damage in metabolic and cardiovascular diseases.

Pathologic heart conditions are characterized by sustained elevation of plasma and interstitial catecholamine levels, as well as by the generation of reactive oxygen species. Endothelial activation is characterized by an abnormal pro-inflammatory. The oxidation of catecholamines also plays an important role. Reactive intermediates formed during catecholamine oxidation have been found to associate with cardiac toxicity.

11.2.1 ROS Generated from Metabolic Bioactivation

Reactive oxygen species are produced by reactive intermediates interacting with oxygen during metabolic activation of foreign compounds such as drugs and other chemicals. Unlike ROS produced in mitochondrial oxidative metabolism, reactive intermediates including reactive oxygen species generated from xenobiotics metabolism play no physiological role. To the contrary, such reactive oxygen species are capable of causing damages to major cellular constituents (proteins, lipids, and DNA).

Metabolic activation of food additives, 2-t-butyl-4-methoxyphenol (BHA) and 2,6-di-t-butyl-4-methylphenol (BHT), to form reactive intermediates may be related to their toxic effects on the lung damage. The oxidation of BHA and BHT catalyzed by horseradish peroxidase and hydrogen peroxide generates unstable phenoxy free radical metabolites.

ROS are more reactive and potent than molecular oxygen itself. Reactive oxygen species include free radicals such as hydroxyl radical (·OH), peroxyl radical (ROO·), alkoxyl radical (RO·), and phenoxy radical ($C_6H_5O^·$). Non-radical reactive oxygen species include hydrogen peroxide (H_2O_2), hydrochlorous acid (HOCl), and super peroxide (O_2^-).

Table 11.1 presents typical reactive oxygen species, including free radicals and non-free radicals, generated from bioactivation of foreign compounds.

Table 11.1 Free radicals and non-free radicals of reactive oxygen species

Free radicals	Non-radicals
Hydroxyl radical (·OH)	Hydrogen peroxide (H_2O_2)
Peroxyl radical (ROO·)	Hydrochlorous acid (HOCl)
Alkoxyl radical (RO·)	Super peroxide (O_2^-)
Phenoxy free radical ($C_6H_5O^·$)	

11.2.2 ROS Generated from Mitochondrial Oxidative Metabolism

Mitochondria carry out important cellular functions including the production of cellular ATP, the regulation of cellular metabolism, and the control of apoptosis. Reactive oxygen species (ROS) produced during mitochondrial oxidative metabolism are referred to as modulators of cellular signaling processes. Mitochondria produced ROS are well recognized for playing a dual role as both deleterious and beneficial species.

The generation of mitochondria-mediated free radicals has been suggested as a necessity for intercellular signaling processes that regulates various physiological functions. At a low or moderate concentration, mitochondria ROS play a physiological role on the defense against infectious agents and a functional role in cellular signaling pathways including signal transduction from membrane receptors. Nevertheless, at a high concentration, mitochondria ROS are hazardous for living organisms and can cause damages to major cellular constituents. Figure 11.1 illustrates mitochondrial oxidative metabolism-generated reactive oxygen species and their association with physiological role and oxidative stress.

11.3 Reactive Nitrogen Species

Metabolic reactive intermediates include not only reactive oxygen species but also reactive nitrogen species. Reactive nitrogen species (RNS), a family of molecules derived from nitric oxide (NO) and nitric superoxide (NO_2), are mainly produced through enzymatic activities involving nitric oxide synthase and NADPH oxidase. Like reactive oxygen species, reactive nitrogen species are also composed of radical species and non-radical species.

Radicals of RNS include nitric oxide (NO) and nitric dioxide (NO_2). Non-radicals of RNS include nitrous oxide (N_2O), peroxynitrite ($ONOO^-$), and derivative of alkyl peroxynitrite (RONOO). Primary reactions of nitric oxide can result in a variety of

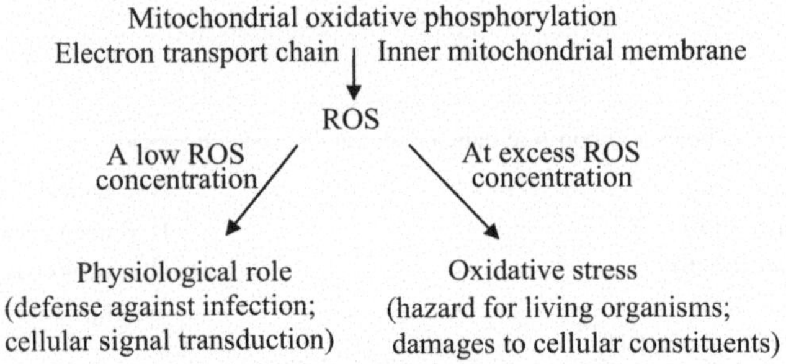

Fig. 11.1 Concentration dependence of mitochondrial reactive oxygen species

Table 11.2 Free radical and non-free radical reactive nitrogen species

Free radicals	Non-radicals
Nitric oxide radical ($^{\bullet}NO^{-}$)	Nitrosonium ion (NO^{+})
Nitrogen dioxide radical ($^{\bullet}NO_2{}^{-}$)	Nitric dioxide (NO_2)
	Nitrous oxide (N_2O)
	Peroxynitrite ($ONOO^{-}$)

secondary products ranging from nitrate (NO^{-3}), nitrite (NO^{-2}), and nitroxyl (NO^{-}) to nitrosonium ion (NO^{+}), peroxynitrite ($ONOO^{-}$), and nitrogen dioxide (NO_2).

Table 11.2 lists typical reactive nitrogen species including free radicals and non-free radicals.

11.4 Oxidative Stress

Oxidative stress is essentially an imbalance between the production of reactive chemical intermediates (including reactive oxygen species and free radicals) and the capability of the body to counteract or detoxify such reactive intermediates through defense mechanisms carried out by antioxidants and detoxification enzymes.

Free radicals affect both the structure and function of neural cells and play the crucial role of oxidative stress in the pathogenesis of neurodegenerative diseases. Reactive oxygen species can cause protein, lipid, and DNA damages. Oxidative stress is considered as a critical mechanism in human pathologies including aging and cancer. Malignant tumors often show increased levels of DNA base oxidation and mutations.

Reactive intermediates of oxygen and nitrogen species (ROS and RNS) may directly oxidize DNA or may interfere with the mechanisms of DNA repair. Inflammatory processes may induce DNA mutations in cells via oxidative stress. This condition occurs when the generation of free radicals and other reactive intermediates exceeds the ability of cells to neutralize and eliminate them. The longer the inflammation persists, the higher the risk of cancer.

Reactive intermediate-mediated oxidative stress is also believed to contribute to the development and progression of a variety of age-related diseases. Reactive intermediates may rapidly react with cell components. For example, 4-hydroxy-2-nonenal, an abundant product of polyunsaturated fatty acid oxidation and decomposition, reacts extensively with DNA and proteins. Parkinson's disease is characterized by a progressive loss of dopaminergic neurons, which is likely associated with dysregulation of oxidation of catechols, such as dopamine and 6-hydroxydopamine, as a result of oxidative stress.

11.5 Oxidative Stress-Mediated Damages

Oxidative stress results from an imbalance between the systemic manifestation of reactive intermediates and reactive oxygen species (ROS) and the capacity of cellular antioxidants and detoxification enzymes to mount an effective antioxidant response. Free radicals with one or more unpaired electrons make them highly reactive with biological molecules, leading to interacting with tissues and cell components.

Overproduction of reactive oxygen molecules and reactive chemical intermediates can cause damages to proteins, lipids, and DNA and impair their functions, thus causing disruptions in normal cellular mechanisms. However, ROS derived from mitochondrial oxidative metabolism are not all harmful to the body. Some of them are useful in killing invading pathogens or microbes.

Figure 11.2 presents reactive intermediate-mediated oxidative stress and the related cellular damages. Oxidative stress-mediated damages include cellular component damages, mitochondrial function intervention, interaction with ion transporters, enzymatic function interference, immune suppression and stimulation effects, and the oxidative degradation of lipids.

Lipid peroxidation is the process in which the unpaired electrons make free radicals highly reactive toward membrane lipids, resulting in lipid peroxidation. The structure and composition of membranes and their susceptibility to peroxidation are dependent on polyunsaturated fatty acid content of the cell membrane. Lipid peroxidation most often affects polyunsaturated fatty acids due to multiple double bonds in between.

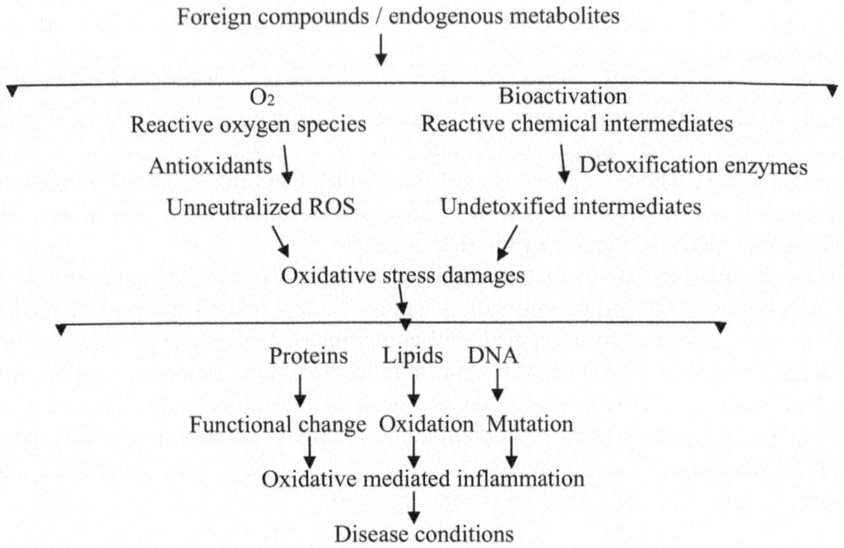

Fig. 11.2 Reactive intermediate-mediated oxidative stress

Reactive oxygen species and reactive chemical intermediates generated from the metabolism of foreign compounds attribute to oxidative stress. Growing evidences have suggested that elevated level of oxidative stress plays a crucial role in the pathogenesis processes contributing to human diseases, such as chronic inflammation, diabetes, neurodegeneration cardiovascular disorders, stroke, rheumatoid arthritis, and cancer.

Inflammation is the body's natural response to a variety of oxidative stress. Oxidative stress-mediated chronic inflammation is a major cause of age-related diseases. Inflammatory process induces a reduction in cellular antioxidant capacity. Uncontrolled inflammatory response can give rise to cell and tissue destruction.

11.6 Defenses Against Oxidative Stress

An imbalance between oxidants and antioxidants is the underlying basis of oxidative stress. The body's antioxidant defense systems against oxidative stress include endogenous antioxidants and exogenous antioxidants. Endogenous antioxidants are products of the body's metabolism, which can be either enzymatic or nonenzymatic. Endogenous antioxidants present in cells are known to act as free radical scavengers, which neutralize free radicals, leading to terminating free radical-generated chain reaction. Endogenous antioxidant enzymes present as a normal defense mechanism of the cells to prevent oxidative stress-mediated damages.

Metabolic detoxification enzymes are endogenous enzymes, which protect against reactive intermediate species and oxygen-derived free radicals. Metabolic syndrome, a risk factors for cardiovascular disease related to oxidative stress, is associated with metabolic syndrome. Oxidative stress, resulting from an antioxidant/prooxidant imbalance, is also crucial in atherogenesis. A weak glutathione-related enzymatic antioxidant was found in human atherosclerotic lesions, suggesting that a specific antioxidant/prooxidant imbalance operative in the vascular wall may be involved in atherogenic processes in humans.

Other important enzymatic antioxidants include catalase, glutathione peroxidase, and glutathione reductase. Investigations of lung antioxidant activities of glutathione peroxidase, superoxide dismutase, catalase, and glutathione S-transferase in different species (rat, hamster, baboon, and human) revealed that marked differences in the sensitivity of various animal species to O_2-induced lung injury may be due in part to variation of cellular antioxidant defenses.

Nuclear factor erythroid 2-related factor 2 (Nrf2) is a transcription factor and regulator of detoxification enzymes. Activation of Nrf2 has been suggested to be an important step in attenuating oxidative stress. The regulation of Nrf2 induction by heme oxygenase-1 was reported to be important for the preferential transcription of detoxification enzymes such as quinone dehydrogenase. Using Nrf2 knockout cells, it was demonstrated that nuclear heme oxygenase-1-associated cytoprotection against oxidative stress depends on the enzyme-Nrf2 interaction.

Nonenzymatic endogenous antioxidants are such as glutathione, uric acid, melatonin, and bilirubin, which are molecules characterized by their ability to rapidly inactivate radicals and oxidants. The body also relies on external sources (exogenous antioxidants) to defend against oxidative stress. Exogenous antioxidants present in food and supplements are essentially non-enzymatic. Phytochemical composition activates Nrf2, induces detoxification enzymes, and protects cardiomyocytes from oxidant-induced apoptosis.

Exogenous antioxidants are present in significant amounts in commonly consumed vegetables and fruits. Consumption of much vegetables and fruits is found to lower the risk of oxidative stress-mediated diseases. Diets contain antioxidants including vitamin E, vitamin C, carotenoids, minerals, and polyphenols. Flavonoids are the most abundant group of polyphenolic compounds found in fruits, vegetables, tea, and wine.

Exogenous antioxidants are also available as dietary supplements such as vitamins, lutein, beta-carotene, and lycopene. Antioxidant vitamins have a number of biological activities such as immune stimulation and prevention of mutation by inhibiting DNA damage induced by reactive oxygen species or metabolites. Alpha-lipoic acid was also reported to reduce oxidative stress and decrease lipid peroxidation. Nevertheless, high-dose supplements of antioxidants may be linked to health issues and may also interfere with medication.

Bibliography

Andersson KE (2018) Oxidative stress and its possible relation to lower urinary tract functional pathology. BJU Int 121(4):527–533

Betteridge DJ (2000) What is oxidative stress? Metabolism 49(2 suppl 1):3–8

Biswas C, Shah N, Muthu M, La P, Fernando AP, Sengupta S, Yang G, Dennery PA (2014) Nuclear heme oxygenase-1 (HO-1) modulates subcellular distribution and activation of Nrf2, impacting metabolic and anti-oxidant defenses. J Biol Chem 289(39):26882–26894

Bolton JL, Dunlap TL, Dietz BM (2018) Formation and biological targets of botanical o-quinones. Food Chem Toxicol 120:700–707

Bryan CL, Jenkinson SG (1987) Species variation in lung antioxidant enzyme activities. J Appl Physiol 63(2):597–602

Chen C-H (2012) Activation and detoxification enzymes: functions and implications. Springer Sciences, New York

Costa VM, Carvalho F, Duarte JA, Bastos ML, Remiao F (2013) The heart as a target for xenobiotic toxicity: the cardiac susceptibility to oxidative stress. Chem Res Toxicol 26(9):1285–1311

Dicker E, Cederbaum AI (1993) Requirement for iron for the production of hydroxyl radicals by rat liver quinone reductase. J Pharmacol Exp Ther 266(3):1282–1290

Drew B, Leeuwenburgh C (2002) Aging and the role of reactive nitrogen species. Ann N Y Acad Sci 959:66–81

Federico A, Morgillo F, Tuccillo C, Ciardiello F, Loguercio C (2007) Chronic inflammation and oxidative stress in human carcinogenesis. Int J Cancer 121(11):2381–2386

Goetz ME, Luch A (2008) Reactive species: a cell damaging rout assisting to chemical carcinogens. Cancer Lett 266(1):73–83

Gower JD (1988) A role for dietary lipids and antioxidants in the activation of carcinogens. Free Radic Biol Med 5(2):95–111

Harman D (1992) Role of free radicals in aging and disease. Ann N Y Acad Sci 673:126–141

Incalza MA, D'Oria R, Natalicchio A, Perrini S, Laviola L, Giorgino F (2018) Oxidative stress and reactive oxygen species in endothelial dysfunction associated with cardiovascular and metabolic diseases. Vascul Pharmacol 100:1–19

Jiang T, Sun Q, Chen S (2016) Oxidative stress: a major pathogenesis and potential therapeutic target of antioxidative agents in Parkinson's disease and Alzheimer's disease. Prog Neurobiol 147:1–19

Jinsmaa Y, Florang VR, Rees JN, Mexas LM, Eckert LL, Allen EM, Anderson DG, Doorn JA (2011) Dopamine-derived biological reactive intermediates and protein modifications: implications for Parkinson's disease. Chem Biol Interact 192(1-2):118–121

Kehrer JP, Klotz LO (2015) Free radicals and related reactive species as mediators of tissue injury and disease: implications for health. Crit Rev Toxicol 45(9):765–798

Lapenna D, de Gioia S, Ciofani G, Mezzetti A, Ucchino S, Calafiore AM, Napolitano AM, Di Ilio C, Cuccurullo F (1998) Glutathione-related antioxidant defenses in human atherosclerotic plaques. Circulation 97(19):1930–1934

Lobo V, Patil A, Phatak A, Chandra N (2010) Free radicals, antioxidants and functional foods: impact on human health. Pharmacogn Rev 4(8):118–126

Meister A, Anderson M (1983) Glutathione. Annu Rev Biochem 52:711–760

Mena S, Ortega A, Estrela JM (2009) Oxidative stress in environmental-induced carcinogenesis. Mutat Res 674(1-2):36–44

McCord JM (2000) The evolution of free radicals and oxidative stress. Am J Med 108:652–659

Milder J, Patel M (2012) Modulation of oxidative stress and mitochondrial function by the ketogenic diet. Epilepsy Res 100(3):295–303

Mironczuk-Chodakowska I, Witkowska AM, Zuiko ME (2018) Endogenous non-enzymatic antioxidants in the human body. Adv Med Sci 63(1):68–78

Morel Y, Barouki R (1999) Repression of gene expression by oxidative stress. Biochem J 342 (Pt 3):481–496

Oyenihi AB, Ayeleso AO, Mukwevho E, Masola B (2015) Antioxidant strategies in the management of diabetic neuropathy. Biomed Res Int 2015:515042

Pearson PG, Soderlund EJ, Dybing E, Nelson SD (1990) Metabolic activation of 1,2-dibromo-3-chloropropane: evidence for the formation of reactive episulfonium ion intermediates. Biochemistry 29(20):4971–4981

Prasad S, Gupta SC, Tyagi AK (2017) Reactive oxygen species (ROS) and cancer: role of antioxidative nutraceuticals. Cancer Lett 387:95–105

Reuland DJ, Khademi S, Castle CJ, Irwin DC, McCord JM, Miller BF, Hamilton KL (2013) Upregulation of phase II enzymes through phytochemical activation of Nrf2 protects cardiomyocytes against oxidant stress. Free Radic Biol Med 56:102–111

Sies H (1997) Oxidative stress: oxidants and antioxidants. Exp Physiol 82:291–295

Simao AN, Dichi JB, Barbosa DS, Cecchini R, Dichi I (2008) Influence of uric acid and gamma-glutamyltransferase on total antioxidant capacity and oxidative stress in patients with metabolic syndrome. Nutrition 24(7-8):675–681

Soloway AH, Curley RW Jr, Soloway SM (2011) Macular degeneration: a possible biochemical mechanism. Med Hypotheses 76(5):729–732

Sun Q, Zhu R, Foss FW Jr, Macdonald TL (2008) In vitro metabolism of a model cyclopropylamine to reactive intermediate: insights into trovafloxacin-induced hepatotoxicity. Chem Res Toxicol 21(3):711–719

Turrens JF (2003) Mitochondrial formation of reactive oxygen species. J Physiol 552:335–344

Tyurina YY, Kapralov AA, Jiang J, Borisenko GG, Potapovich AL, Sorokin A, Kochanek PM, Graham SH, Schor NF, Kagan VE (2006) Oxidation and cytotoxicity of 6-OHDA are mediated by reactive intermediates of COX-2 overexpressed in PC12 cells. Brain Res 1093(1):71–82

Valoti M, Sipe HJ Jr, Sgaragli G, Mason RP (1989) Free radical intermediates during peroxidase oxidation of 2-t-butyl-4-methoxyphenol, 2,6-di-t-butyl-4-methylphenol, and related phenol compounds. Arch Biochem Biophys 269(2):423–432

West JD, Marnett LJ (2005) Alterations in gene expression induced by the lipid peroxidation product, 4-hydroxy-2-nonenal. Chem Res Toxicol 18(11):1642–1653

Chapter 12
Xenobiotic Bioactivation-Mediated Cellular Damages

Natural toxins are toxic compounds that are produced by living organisms. Although the risk for toxicity due to consumption of food toxins is low, there is always the possibility of toxicity due to contamination, overconsumption, allergy, or an unpredictable individual response. Natural toxins can cause a variety of adverse health effects, which can pose a serious health threat to humans. Long-term health effects of natural toxicity include damages to immune, nervous, or reproductive system, leading to various disease conditions.

Besides natural toxins, many toxicities are resulted from the metabolic activation of foreign compounds. Reactive intermediates and metabolites derived from foreign compound metabolism are potentially capable of interacting with endogenous targets, which can trigger perturbation in cellular functions and result in cellular damages. Initially, the cells respond to such perturbation with repair and adaptive mechanisms. But, when the induced perturbation exceeds the repair and adaptive capacity, toxic effects occur. In most cases, foreign compounds are absorbed and distributed to target organs, where harmful effects occur.

Covalent binding of reactive intermediates to cellular components (proteins, lipids, and DNA) results in forming protein and DNA adducts and causing lipid peroxidation. The liver and kidney are frequent target organs of toxicity. The extent of foreign compound toxicity depends on its nature, metabolic enzymes, and the environmental factors. Extensive research has been carried out on electrophilic reactive intermediates interacting with cellular nucleophilic sites or residues. Another important research area having also been carried out is the adverse drug reactions, due to the generation of toxic reactive intermediate in the course of drug metabolism.

C.-H. Chen, *Xenobiotic Metabolic Enzymes: Bioactivation and Antioxidant Defense*,
https://doi.org/10.1007/978-3-030-41679-9_12

12.1 Exhibition of Foreign Compound Toxicities

The exhibition of foreign compound toxicity occurs mainly under the following three circumstances: (a) the presence of natural toxicity in the parent compounds like toxigenic fungi and toxic chemicals; (b) metabolic activation that generates reactive intermediates which ultimately exhibits toxicity, including reactive oxygen/nitrogen species and free radicals; and (c) the presence of induced toxicity through modulation by the environmental factors such as smoke and alcohol. These three circumstances are discussed below:

12.1.1 Natural Toxicity

Toxigenic fungi exist in nature and occur frequently in food supplies due to mold infestation of susceptible agricultural products. When toxigenic fungi are present in foods in sufficiently high levels, fungal metabolites can have toxic effects ranging from acute (liver or kidney deterioration) to chronic disease conditions (such as liver cancer). Other foreign compounds that exhibit natural toxicity include nicotine, heavy-metal ions, and carbon monoxide.

Nicotine affects cell proliferation, oxidative stress, and apoptosis, resulting in mediating health hazards including an increasing risk of cardiovascular, respiratory, and gastrointestinal disorders, as well as a decrease in immune response. Exposure to heavy metals such as cadmium, mercury, and lead can result in severe health hazards such as prenatal and developmental defects. Carbon monoxide binding to myoglobin causes ineffectiveness in delivering oxygen to body tissues, leading to significant damages to the heart and central nervous system.

12.1.2 Bioactivation-Mediated Toxicity

Metabolic activation of foreign compounds, such as foods, drugs, and other chemicals, to produce reactive intermediates, plays an important role in cytotoxicity and carcinogenesis. Activation enzymes catalyze functionalization reactions that convert foreign compounds to water-soluble metabolites, which are followed by conjugation reactions catalyzed by detoxification enzymes to form less toxic or inactive metabolites, thus facilitating the excretion of foreign compounds from the body. However, during the course of xenobiotic metabolism, in many cases, potentially toxic reactive intermediates and metabolites are formed.

Table 12.1 briefly presents the exhibition of toxicity of foreign compounds, including natural toxicity, bioactivation-mediated toxicity, and induced toxicity.

Table 12.1 Sources of foreign compound toxicity

Foreign compound toxicity	Sources
Natural toxicity	Fungi; heavy metals (cadmium, mercury, and lead); nicotine; carbon monoxide
Bioactivation-mediated toxicity	Reactive intermediates: reactive oxygen/nitrogen species and free radicals
Induced toxicity	Alcohol; smoke

12.1.2.1 Adverse Drug Reactions

Adverse drug reactions are resulted from generating toxic reactive intermediates and metabolites during drug metabolism. For example, CYP450 enzymes are responsible for metabolism of acetaminophen, a primary ingredient in cold and pain relief remedies. CYP450-catalyzed reactions metabolize acetaminophen to produce reactive intermediates, quinoneimine and semiquinoneimine. Quinone compounds exhibit toxic effects by binding to hepatic and renal proteins. An overdose of acetaminophen can trigger potential liver damage, resulting in major acute liver failure.

12.1.2.2 Carcinogens

Chemical carcinogens belong to a diverse group of compounds. A variety of human carcinogens have been reported, including 2-naphthylamine, aflatoxins, arsenic and nickel compounds, asbestos, benzene, benzidine, tamoxifen, tobacco smoke, and polychlorinated biphenyl. Other carcinogens take effect after chemicals have been converted to toxic reactive intermediates or metabolites catalyzed by metabolic activation enzymes.

Some of carcinogens act on the target cells, and a majority of them are mutagenic. Mutagenic chemicals are DNA-damaging agents, which are considered to produce enduring alterations in the genetic substances in the cells. Among DNA-damaging agents, there are direct-acting carcinogens and indirect-acting carcinogens. Direct-acting carcinogens, such as N-nitroso-N-methylurea and methyl methanesulfonate, are compounds that are intrinsically reactive and can covalently interact with DNA to form DNA adducts without requiring bioactivation,

Indirect-acting carcinogens require metabolic activation of parent compounds to produce reactive intermediates, such as reactive oxygen/nitrogen species or free radicals, which can covalently bind to DNA to form DNA adducts. Reactive intermediates produced during metabolic activation are generally electrophilic metabolites that are capable of interacting with nucleophilic sites in DNA.

12.1.3 *Induced Toxicity*

The exposure to alcohol enhances the vulnerability of hepatotoxicity. Alcohol can activate CYP2E1 isozyme that catalyzes the activation of acetaminophen. Chronic ethanol consumption can potentiate acetaminophen hepatotoxicity through enhancing N-acetyl-p-benzoquinone imine formation. Studies of the effects of chronic ethanol consumption have revealed that acetaminophen toxicity, as manifested by liver enlargement and congestion, is significantly increased. The enhanced toxicity is a result of alcohol's effects on hepatocyte membranes, which renders the cells more susceptible to toxic injury.

Tobacco smoke is another example of induced toxicity. The effects of tobacco smoke on the respiratory system include reduction of lung function, impairment of lung clearance system, increasing risk of the lung infection, and damage to the air sacs of lungs. Tobacco smoke also affects the circulatory system including raising blood pressure and heart rate and increasing risk of stroke and heart attack. Tobacco smoke also affects the immune system, leading to greater susceptibility to pneumonia and influenza infections.

12.2 Oxidative Stress Generated from Xenobiotic Metabolism

Metabolic activation of foreign compounds requires functionalization reactions catalyzed by activation enzymes such as CYP450, flavin-monooxygenase, lipoxygenase, and xanthine oxidase. Chemically active intermediates react with an oxygen molecule to form reactive oxygen species. Nitric oxide radical is generated enzymatically by nitric oxide synthase from L-arginine. Such reactive chemical intermediates including reactive oxygen and nitrogen species, and free radicals are major contributors to oxidative stress.

Oxidation stress is among the fundamental mechanisms that contribute to the toxicity of many foreign compounds. Oxidative stress can make biological systems unable to repair the resulting damages, although conjugation reactions catalyzed by detoxification enzymes serve as a defense system to detoxify reactive intermediates. However, an imbalance occurs if the rate of producing reactive intermediates is significantly higher than that of detoxifying reactions or if the overly generated reactive intermediates are not compensated by the body's antioxidant defense systems.

12.3 Interactions with Cellular Components

When produced intracellularly in sufficient quantities, reactive intermediates, particularly reactive oxygen species and free radicals, have the capacity to interact with cellular components (proteins, DNA, and lipids), ultimately leading to alterations in the structure and function of target cellular molecules. The interactions of chemically active intermediates with cellular components can be either specific or non-specific. Specific interaction involves binding to a specific site in a receptor, enzyme, or transport protein.

Figure 12.1 illustrates the interactions of chemically active intermediates with proteins, lipids, and DNA, leading to the formation of protein adducts, DNA adducts, and lipid peroxidation. The figure reveals the potential toxic effects as a result of interactions with cellular components. Such toxic effects could contribute to a variety of disease conditions, such as aging, cardiovascular disease, neurological disorder, and cancer.

Cellular molecules are major targets for oxidative damages. Foreign compound-mediated oxidative protein damage can cause an impairment of enzyme catalytic functions or other protein functions. Oxidative DNA damage can affect DNA base pairing, causing a mismatch in the DNA base pair transformation and resulting in a mutation. Oxidation of cell membranes can lead to peroxidation of membrane lipid, which is a primary mechanism of tissue injury.

The liver is not only the principal site for the activation of foreign compounds, but also the center of production of reactive intermediates, and the primary site of toxic effects. However, chemically active intermediates may also be exported to other tissues and cause harmful effects there, such as in the case of aromatic amines which are potential urinary bladder carcinogens.

12.3.1 *Interactions with Proteins*

Reactive intermediates mediated by foreign compounds are able to interact with cellular proteins, resulting in oxidative protein damage. If the oxidative damage is not repaired, loss of catalytic function for enzymes and impairment of protein function can occur. Among amino acid residues, the sulfur-containing amino acids, cysteine and methionine, are particularly susceptible to oxidation by reactive

$$\text{Reactive intermediate} \xrightarrow{\text{Protein}} \text{Protein adduct} \rightarrow \text{Cytotoxicity}$$

$$\text{Reactive intermediate} \xrightarrow{\text{DNA}} \text{DNA adduct} \rightarrow \text{Mutation} \rightarrow \text{Malignancy}$$

$$\text{Reactive oxygen species} \xrightarrow{\text{Lipid}} \text{Lipid peroxidation} \rightarrow \text{Tissue damages}$$

Fig. 12.1 Interactions of reactive intermediates with cellular molecules

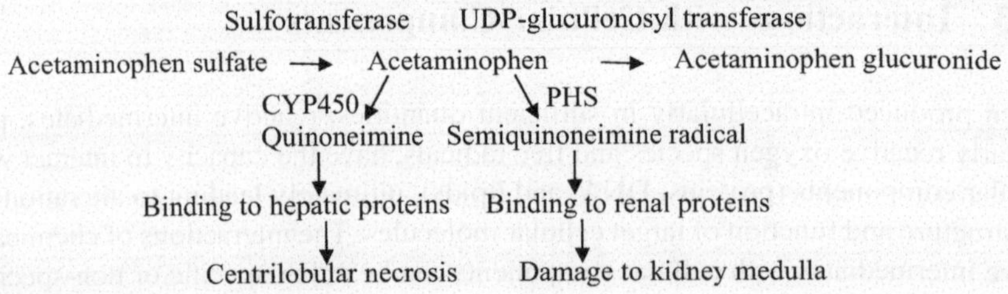

Fig. 12.2 Metabolic intermediate acetaminophen binding to protein

oxygen species. Hemoglobin is also a frequent target for oxidative damage. Xenobiotic-induced oxidation of two globin chains of hemoglobin can lead to disulfide bond formation, resulting in the interference of its folding and the loss of its function.

Besides sulfur-containing amino acid residues (cysteine and methionine), the primary binding sites of protein adducts include nitrogen of lysine or histidine residues. Although not all metabolic intermediate bindings to protein are toxic in nature, protein adducts are the source of toxicity mediated by foreign compounds. The binding can result in inactivation of enzymatic functions, as in the case of microcystin binding to phosphatase, which leads to the inactivation of the enzyme.

Using drug acetaminophen as an example, Fig. 12.2 illustrates how metabolic intermediate binding to protein leads to liver and kidney damages.

12.3.2 Interactions with DNA

Reactive chemical intermediates, including reactive oxygen/nitrogen species and free radicals, are capable of causing DNA damage, which eventually may attribute to cell transformation and tumor initiation. The oxidation of a DNA base by reactive oxygen species has severe consequences on DNA base pairing, such as the oxidation at the guanine residue resulting in the insertion of a wrong base (adenine) in DNA. If the change in base pairing persists without repair during the next replication cycle, DNA damage can occur.

Reactive metallic intermediates are also one group of foreign compounds that are associated with oxidative DNA damage. Nickel sulfides and other nickel compounds can give rise to the production of highly reactive hydroxyl radicals that can attack DNA bases, causing oxidative DNA damage. Binding a nickel compound to DNA can lead to an increased risk of respiratory cancer.

Any of four bases in DNA may participate in the interaction of reactive intermediate with DNA base. Among them, guanine residue is the predominant base for the binding. Electron-dense sites, like nitrogen and oxygen atoms in nucleic acid, are primary binding sites. DNA adduct formation can affect DNA base pairing. When

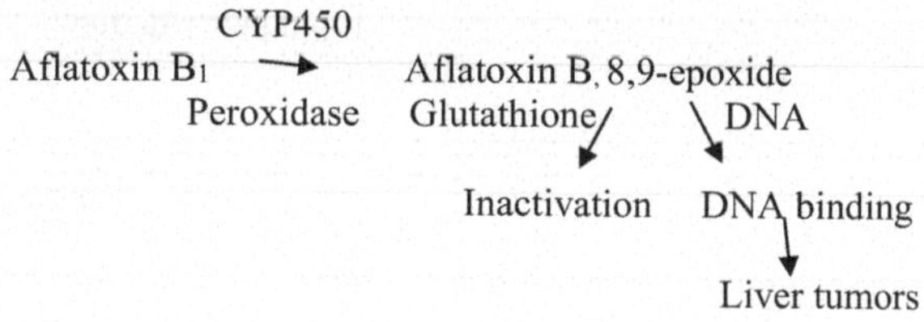

Fig. 12.3 Metabolic intermediate aflatoxin B_1 binding to DNA

DNA undergoes replication, erroneous base pairing can result in base mismatch. Mutation can occur if base mismatch is not repaired during DNA replication.

Using aflatoxin B_1 as an example, Fig. 12.3 illustrates how metabolic intermediate of aflatoxin B_1 binding to DNA can result in liver damages.

12.3.3 *Interactions with Lipids*

Membrane lipid peroxidation has been implicated as one of the mechanisms that cause xenobiotic-induced tissue injury. Lipid peroxidation refers to the oxidative degradation of membrane lipids. When reactive oxygen species or free radicals are generated in the membrane compartment of the cell, oxidation of membrane lipids occurs, which can lead to membrane peroxidation and oxidative lipid damage. For instance, trichloromethyl radical produced during metabolic activation of carbon tetrachloride catalyzed by CYP450 enzyme can trigger membrane lipid peroxidation.

Lipid peroxidation involves the oxidation of fatty acyl carbon chains and the formation of lipid peroxyl radicals. The peroxidation reaction also produces reactive aldehyde that can expand the toxicity of foreign compounds. Lipid peroxidation most often affects polyunsaturated fatty acids that contain multiple double bonds between methylene groups and possess reactive hydrogens. Lipid peroxidation is a propagating chain reaction resulting in damaging lipid bilayers, the backbone of cell membranes.

The mechanism of lipid peroxidation consists of three major steps: initiation, propagation, and termination. Initially, fatty acid free radical is produced when reactive oxygen species such as OH· reacts with unsaturated lipid. The resulting unstable fatty acid free radical can react with oxygen to yield peroxyl-fatty acid radical, which can then react with another fatty acid free radical to produce a different fatty acid radical and lipid peroxide. Figure 12.4 briefly describes lipid peroxidation and oxidative degradation of membrane lipids.

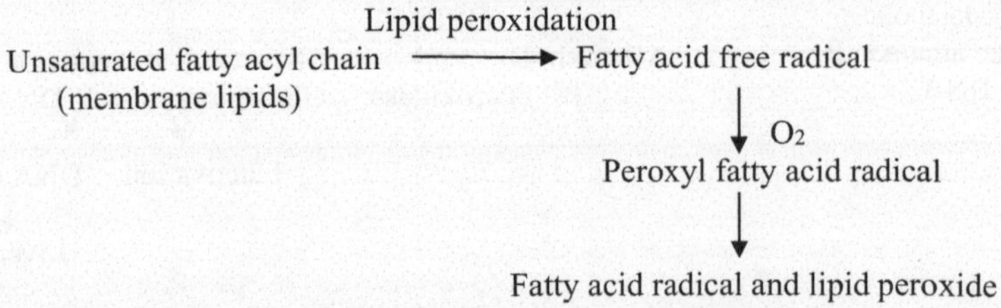

Fig. 12.4 Lipid peroxidation and oxidative degradation of membrane lipids

Table 12.2 Foreign compound-mediated cellular functional damages

Effects on cell functions	Functional damages
Interaction with cellular components	Impairment in protein functions; DNA base pair mismatch and mutation; lipid peroxidation; tissue damages
Mitochondrial function intervention	Inhibition of electron transport chain, impairment of energy supply; inhibition of ATP synthesis
Interaction with ion transporters	Interference of channel or pump; inhibition of Na^+, K^+ pump; inhibition of sodium channel
Enzymatic function interference	Inhibition of enzyme activity; enzymatic inactivation; interacting with antioxidant enzymes
Immune suppression and stimulation effects	Repression of immune cell development; enhancement in immune response; tissue damage and immune-mediated disease

12.4 Cellular Functional Damages

Metabolic reactive intermediate-mediated oxidative stress affects many important cellular functions, including (a) mitochondrial function intervention, (b) interaction with ion transporters, (c) enzymatic function interference, and (d) immune suppression and stimulation effects. Table 12.2 lists these foreign compound-mediated cellular functional damages.

12.4.1 Mitochondrial Function Intervention

Mitochondria are crucial to the maintenance of many cellular functions, particularly energy utilization. The production of ATP is central to the supply of energy needed for biosynthesis and transport processes. Inhibition of mitochondrial functions subsequently causes a decrease in the production of cellular ATP, leading to an impairment of energy supply. Foreign compound-mediated oxidative stress could impair mitochondria functions through their interference with electron transport chains or by inhibiting ATP synthetase activity.

Industrial chemical pentachlorophenol, a wood preservative, is endogenously formed by metabolism of hexachlorobenzene. Pentachlorophenol is absorbed into the body by occupational exposure, causing toxic effects through its ability to uncouple oxidative phosphorylation of liver mitochondria. Mitochondrial dysfunction has also been implicated in the pathogenesis of valproic acid-mediated hepatotoxicity. Valproic acid is an antiepileptic drug. Acute treatment of freshly isolated rat hepatocytes with valproic acid was found to result in oxidative stress, while glutathione provides protection to hepatocytes against mitochondrial damage by valproic acid.

Rotenone, another foreign compound, is used as an insecticide and pesticide. It occurs naturally in the roots and stems of plants such as vine plants. Rotenone is capable of intervening with mitochondria function. It is an inhibitor of oxidative phosphorylation, since it is able to block electron transfer in NADH-dependent oxidoreductase and prevent the utilization of NADH as a substrate.

Inhibition of the electron transport chain obstructs ATP synthesis. In the presence of uncouples, the proton motive force across mitochondria membranes is dissipated, and ATP is not formed by mitochondrial ATP synthase. Chemical compound 2,4-dinitrophenol can cause uncoupling of electron transport and affect phosphorylation in mitochondria. 2,4-dinitrophenol was found to have the ability to greatly increase metabolic rate, causing the rapid loss of body fat. However, concerns about its side effects prevent its use for weight loss purposes.

12.4.2 *Interactions with Ion Transporters*

Ion transporters like channels and pumps have the capability of transporting specific ions across biological membranes. Na^+, K^+ pump is an ATP-dependent transporter which mediates the efflux of Na^+ in exchange of influx of K^+, leading to building up a Na^+ gradient across cell membranes with a higher extracellular Na^+. The generated Na^+ gradient can drive Na^+-dependent secondary transport. Interference of a channel or a pump at its binding site by a foreign compound may inhibit the capacity of an ion transporter.

In cardiomyocytes, the influx of extracellular Na^+ into the cell drives the efflux of Ca^{2+} from the cell. The concentration of intracellular Ca^{2+} appears to associate with contraction in cardiac muscle cells. Digitoxin, a cardiac glycoside, mediates the inhibition of Na^+, K^+ pump. Its binding to the outer surface of ATPase blocks Na^+ and K^+ translocation events as well as ATP hydrolysis. Specific inhibition of Na^+, K^+ pump by digitoxin results in an increase in the intracellular Ca^{2+} levels, which appears to stimulate contraction in cardiac muscle cells. Digitalis is used in treating cardiac conditions such as congestive heart failure. However, an overdose of digitoxin can cause adverse drug reactions. Over stimulation of cardiac muscle cells can lead to cardiac arrhythmia.

Tetrodotoxin, a neurotoxin, is another example of foreign compound intervention with ion transporter. Tetrodotoxin specifically binds and inhibits a sodium channel in excitable nerve tissues. It blocks action potentials in nerves by binding to the pores of the voltage-dependent sodium channels in nerve cell membranes, thus essentially preventing the affected nerve cells from firing. Tetrodotoxin therefore is a potent neurotoxin. It is present in the organs of puffer fish. Poisoning with tetrodotoxin can occur after the ingestion of puffer fish.

12.4.3 Interference with Enzymatic Functions

Interference with enzymatic functions refers to certain conditions when the enzyme cannot carry out its normal function as the catalyst for a specific reaction. When inhibited by a toxic foreign compound, an enzyme may no longer carry out the conversion of the substrate. Many enzymes and proteins contain a sulfhydryl group at their catalytic site, which are vulnerable to thiol-reactive compounds.

Heavy metals exhibit high affinity for the thiol group. Arsenic has been known to selectively inhibit pyruvate dehydrogenase, an enzyme complex which links the glycolytic pathway with the citrate cycle. The binding of non-metallic compounds at a catalytic site may also cause enzyme inactivation.

Enzymatic functions can also be interfered by environmental factors such as the chronic consumption of alcohol and exposure to environmental chemicals. Enzyme CYP2E1 contributes significantly to ethanol metabolism and formation of acetaldehyde, a highly reactive metabolite. CYP2E1 is hence involved in oxidative stress produced by ethanol.

Organophosphates are important in agricultural industries as pesticides to protect plants. They are lipophilic compounds readily absorbed and distributed in the body. The binding of an organophosphate to acetylcholinesterase inhibits its enzymatic activity. Some organophosphates were shown to exhibit acute neurotoxicity. Acetylcholinesterase is a serine esterase whose function is to hydrolyze the neurotransmitter acetylcholine. The serine hydroxyl group is normally attacked by the substrate acetylcholine. However, in the presence of an organophosphate, the active serine hydroxyl group of acetylcholinesterase is attacked by the phosphorus of the organophosphate, resulting in the inhibition of acetylcholinesterase activity.

A number of potential toxic foreign compounds and their metabolites are also able to weaken the body's antioxidant defense by interacting with antioxidants and antioxidant enzymes. Antioxidants protect cells against adverse effects of oxidants. Antioxidant enzymes provide scavengers for reactive oxygen/nitrogen species, thus serving as a brake to prevent over presence of these species, such as superoxide dismutase which scavenges superoxide free radical by converting it to peroxide, which is then destroyed by catalase.

12.4.4 Immune Suppression and Stimulation Effects

Repression of an immune cell mediated by reactive intermediates of foreign compounds may lead to immune suppression. Suppression of the immune system renders an organism susceptible to infections, compromising the ability of the body to fight diseases. Side effects of immunosuppressive agents also include hypertension, peptic ulcers, and liver and kidney injury. Immunosuppressive agents are drugs that inhibit or reduce the activation or efficacy of the immune system. Potential immune suppressants include steroids such as cortisone.

In contrast to an immunosuppressive effect, some reactive intermediates or metabolites can cause an enhancement in immune response, leading to tissue damage and immune-mediated disease. Chemical allergy is initiated by the immune system. Penicillin is the most common drug that can draw an allergic response to a minority of recipients. Over stimulation of the immune system may also result in hyposensitivity to immunoallergic reaction.

Enhancement of the immune system can lead to chemical hypersensitivity. Such an adverse reaction to a reactive chemical is the result of previous sensitization to that chemical or its structurally similar compound. After an initial allergic reaction to that chemical, a small subsequent exposure can evoke a much more severe response.

Environmental pollutant such as dioxin has the capacity of eliminating maturing T cells in the thymus. Moreover, direct addition of alkylators, such as N-nitroso-N-methylurea and dimethyl sulfate, to naive murine splenocytes has been reported to produce a dose-dependent suppression of the in vitro antibody-forming cell in response to antigens.

12.5 Attributing to Various Disease Conditions

Oxidative stress arises from an imbalance between the production of reactive intermediates, including reactive oxygen/nitrogen species and free radicals, and the capability of biological system to readily detoxify them. Such an imbalance represents the primary cause of cellular stress and dysfunction. Reactive and nitrogen species are known to contribute as pathogenic factors to the development of chronic progressive diseases. Reactive intermediate-mediated oxidative stress is believed to attribute to the development and progression of a variety of age-related diseases such as neurodegeneration.

Aging is the progressive loss of tissue and organ functions. The oxidative stress theory of aging hypothesizes that age-associated functional losses are due to the accumulation of oxidative damages to cellular molecules including proteins, lipids, and DNA. Neural cell functional damages can result in age-related neurodegenerative diseases.

Oxidative cell damages have also been implicated in many other disease conditions such as cardiovascular risk, hypertension, metabolic disorders, and cancer. Oxidative stress-mediated inflammation and cellular damages are believed to be linked to cardiovascular diseases. Cardiovascular dysfunction is leading cause for the mortality of diabetic individuals. Antioxidant agents were found to be helpful in the prevention of diabetes-induced cardiovascular dysfunction.

DNA mutation is a critical step in carcinogenesis, and oxidative stress is involved in the carcinogenesis process. Continued oxidative stress leads to chronic inflammation, which in turn could mediate many chronic diseases including cancer. Oxidative damage mediated by reactive intermediate species such as hydroxyl radical ($OH^{\bullet}$) are capable of damaging DNA and promoting the development of malignancy and cancer. Oxidative stress-induced cellular redox imbalance has been shown in cancer cells compared with normal cells. Such imbalance may be related to oncogenic stimulation.

12.6 Mediation of Adverse Drug Reactions

Adverse drug reactions are a public health problem. Some drug metabolism can attribute to cytotoxicity through the bioactivation to form chemically reactive intermediate. Adverse drug reactions such as hepatotoxicity are a major obstacle for the development of medicines. The toxicities of hepatotoxins were found to correlate with the generation of reactive intermediates and metabolites. The covalent binding of reactive drug intermediate and the formation of drug-protein adducts may result in direct cell toxicity. By modifying cellular macromolecules, reactive intermediates and metabolites play a pivotal role in the pathogenesis of adverse drug reactions.

Pain medicine acetaminophen has been exemplified for the underlying mechanisms of reactive intermediate-mediated hepatotoxicity. Acetaminophen is metabolized in the liver to N-acetyl-p-benzoquinone imine, an electrophilic metabolite known to bind liver proteins, resulting in hepatotoxicity. Reactive intermediates of furosemide, a diuretic drug, also cause acute liver injury, leading to the investigation of antioxidant protection against furosemide-induced liver injury. Moreover, bioactivation of trimethoprim, an antibacterial agent, to reactive iminoquinone methide intermediate has been proposed as a cause for its toxicity. Thalidomide, a sedative and antiemetic drug, is bioactivated to arene oxide and quinone reactive intermediates, which are responsible for the toxicity of thalidomide.

Bibliography

Arispe N, Diaz JC, Simakova O et al (2008) Heart failure drug digitoxin induces calcium uptake into cells by forming transmembrane calcium channels. Proc Natl Acad Sci U S A 105:2610–2615

Blaikie FH, Brown SE, Samuelsson LM et al (2006) Targeting dinitrophenol to mitochondria: limitations to the development of a self-limiting mitochondrial protonophore. Biosci Rep 26:231–243

Brzezinski MR, Boutelet-Bochan H, Person RE et al (1999) Catalytic activity and quantitation of cytochrome P-450 2E1 in prenatal human brain. J Pharmacol Exp Ther 289:1648–1653

Chen CH (2012) Activation and detoxification enzymes: functions and implications. Springer Sciences, New York

Choi DW, Leininger-Muller B, Wellman M et al (2004) Cytochrome p-450-mediated differential oxidative modification of proteins: albumin, apolipoprotein E, and CYP2E1 as targets. J Toxicol Environ Health A 67:2061–2071
Chou AP, Li S, Fitzmaurice AG et al (2010) Mechanisms of rotenone-induced proteasome inhibition. Neurotoxicology 31:367–372
Damsten MC, de Vlieger JS, Niessen WM, Irth H, Vermeulen NP, Commandeur JN (2008) Trimethoprim: novel reactive intermediates and bioactivation pathways by cytochrome p450s. Chem Res Toxicol 21(11):2181–2187
Dolan LC, Matulka RA, Burdock GA (2010) Naturally occurring food toxins. Toxins (Basel) 2 (9):2289–2332
Driscoll JP, Yadav AS, Shah NR (2018) Role of glucuronidation and P450 oxidation in the bioactivation of Bromfenac. Chem Res Toxicol 31(4):223–230
Fleming TH, Humpert PM, Nawroth PP, Bierhaus A (2011) Reactive metabolites and AGE/RAGE-mediated cellular dysfunction affect the aging process: a mini-review. Gerontology 57 (5):435–443
Goetz ME, Luch A (2008) Reactive species: a cell damaging rout assisting to chemical carcinogens. Cancer Lett 266:73–83
Goldman JL, Koen YM, Rogers SA, Li K, Leeder JS, Hanzlik RP (2016) Bioactivation of trimethoprim to protein-reactive metabolites in human liver microsomes. Drug Metab Dispos 44(10):1603–1607
Gonzalez FJ (2005) Role of cytochromes P450 in chemical toxicity and oxidative stress: studies with CYP2E1. Mutat Res 569:101–110
Goyer RA, Clarkson TW (2001) Toxic effects of metals. In: Klaasen CD (ed) Casarett and Doullis toxicology: the basic science of poisons, 6th edn. McGraw-Hill, New York, pp 861–867
Haggerty HG, Kim BS, Holsapple MP (1990) Characterization of the effects of direct alkylators on in vitro immune responses. Mutat Res 242:67–78
Halliwell B (2007) Oxidative stress and cancer: have we moved forward? Biochem J 401(1):1–11
Hodgon E, Smart RC (2001) Introduction to biochemical toxicology. Wiley, New York
Incalza MA, D'Oria R, Natalicchio A, Perrini S, Laviola L, Giorgino F (2018) Oxidative stress and reactive oxygen species in endothelial dysfunction associated with cardiovascular and metabolic diseases. Vascul Pharmacol 100:1–19
Jaeschke H, Gores GJ, Cederbaum AI et al (2002) Mechanisms of hepatotoxicity. Toxicol Sci 65:166–176
Jan YH, Heck DE, Dragomir AC, Gardner CR, Laskin DL, Laskin JD (2014) Acetaminophen reactive intermediates target hepatic thioredoxin reductase. Chem Res Toxicol 27(5):882–894
Liebler DC, Guengerich FP (2005) Elucidating mechanisms of drug-induced toxicity. Nat Rev Drug Discov 4(5):410–420
Liguori I, Russo G et al (2018) Oxidative stress, aging, and diseases. Clin Interv Aging 13:757–772
Luís PB, Ruiter JP, Aires CC et al (2007) Valproic acid metabolites inhibit dihydrolipoyl dehydrogenase activity leading to impaired 2-oxoglutarate-driven oxidative phosphorylation. Biochim Biophys Acta 1767:1126–1133
Masini A, Ceccarelli-Stanzani D et al (1985) The role of pentachlorophenol in causing mitochondrial derangement in hexachlorobenzene induced experimental porphyzria. Biochem Pharmacol 34:1171–1174
Mates JM (2000) Effects of antioxidant enzymes in the molecular control of reactive oxygen species toxicology. Toxicology 153:83–104
Pamplona R (2008) Membrane phospholipids, lipoxidative damage and molecular integrity: a causal role in aging and longevity. Biochim Biophys Acta 1777:1249–1262
Perez MJ, Cederbaum AI (2003) Adenovirus-mediated expression of Cu/Zn- or Mn-superoxide dismutase protects against CYP2E1-dependent toxicity. Hepatology 38:1146–1158
Qu Q, Liu J, Zhou HH, Klaassen CD (2014) Nrf2 protects against furosemide-induced hepatotoxicity. Toxicology 324:35–42

Spracklin DK, Hankins DC, Fisher JM et al (1997) Cytochrome P450 2E1 is the principal catalyst of human oxidative halothane metabolism in vitro. J Pharmacol Exp Ther 281:400–411

Thompson RA, Isin EM, Ogese MO, Mettetal JT, Williams DP (2016) Reactive metabolites: current and emerging risk and hazard assessments. Chem Res Toxicol 29(4):505–533

Tong V, Teng XW, Chang TK, Abbott FS (2005) Valproic acid II: effects on oxidative stress, mitochondrial membrane potential, and cytotoxicity in glutathione-depleted rat hepatocytes. Toxicol Sci 86:436–443

Turan B (2010) Role of antioxidants in redox regulation of diabetic cardiovascular complications. Curr Pharm Biotechnol 11(8):819–836

Uttara B, Singh AV, Zamboni P, Mahajan RT (2009) Oxidative stress and neurodegenerative diseases: a review of upstream and downstream antioxidant therapeutic options. Curr Neuropharmacol 7(1):65–74

Valko M, Rhodes CJ, Moncol J, Izakovic M, Mazur IM (2006) Free radicals, metals and antioxidants in oxidative stress-induced cancer. Chem Biol Interact 160(1):1–40

Vassort G, Turan B (2010) Protective role of antioxidants in diabetes-induced cardiac dysfunction. Cardiovasc Toxicol 10(2):73–86

Wani TH, Chakrabarty A, Shibata N, Yamazaki H, Guengerich FP, Chowdhury G (2017) The dihydroxy metabolite of the teratogen thalidomide causes oxidative DNA damage. Chem Res Toxicol 30(8):1622–1628

Wells PG, Kim PM, Laposa RR et al (1997) Oxidative damage in chemical teratogenesis. Mutat Res 396:65–78

West JD, Marnett LJ (2005) Alterations in gene expression induced by the lipid peroxidation product, 4-hydroxy-2-nonenal. Chem Res Toxicol 18:1642–1653

Chapter 13
Nrf2-ARE Pathway: Defense Against Oxidative Stress

Metabolic conversions of foreign compounds catalyzed by metabolic enzymes, in many cases, lead to the formation of reactive chemical intermediates including reactive oxygen species and free radicals. Such chemical reactive species can cause cell damages by attacking cellular components (proteins, membrane lipids, and DNA), leading to cytotoxicity, lipid peroxidation, and carcinogenesis. The body utilizes antioxidative stress defense systems in the cells to neutralize or eliminate these chemical reactive species.

Nevertheless, an imbalance occurs when the production of reactive chemical intermediates exceeds the ability of metabolic detoxification enzymes and endogenous antioxidants to neutralize them. This imbalance can lead to oxidative stress resulting in cellular damages. To maintain a balance between bioactivation and detoxification, living cells maintain an inducible antioxidant response pathway, which activity is largely regulated by nuclear factor erythroid-2-related factor 2 (Nrf2). It should be pointed out that growing evidence indicates that other molecular mechanisms may also be involved.

Nrf2 governs the expression of many detoxification enzymes and antioxidants and acts as a sensor of oxidative stress and prevents genome instability. Nrf2 has become an important chemopreventive and therapeutic target. Nrf2 pathway is the primary mechanism of Nrf2 activation, which is based on dissociation of Nrf2 from its inactive complex with the repressor protein Keap1 and the subsequent translocation of Nrf2 into the nucleus.

During oxidative stress, the Nrf2–Keap1 interaction is disrupted, permitting Nrf2 to translocate to the nucleus. Once in the nucleus, Nrf2 complex binds to antioxidant response elements (ARE), promoting transcription of cytoprotective genes. Nrf2-ARE-dependent genes include NAD(P)H:quinone oxidoreductase, glutathione *S*-transferase, UDP-glucuronosyltransferase, and other detoxification enzymes.

C.-H. Chen, *Xenobiotic Metabolic Enzymes: Bioactivation and Antioxidant Defense*,
https://doi.org/10.1007/978-3-030-41679-9_13

13.1 Oxidative Stress

Oxidative stress is a biochemical dysregulation, while inflammation is a biological response. Chronic inflammation is strongly associated with the major stages of cytotoxicity and carcinogenesis. The cellular damages resulting from oxidative stress are a fundamental pathological process in a variety of diseases including aging and cancer.

Oxidative stress associated cellular damages, such as protein adducts, lipid peroxidation, and DNA mutagenicity, can contribute to the progression of aging symptoms and many disease conditions. For aging-related symptoms, there are evidences pointing to oxidative stress in association with chronic neuroinflammation as a pathological mechanism underlying neurodegeneration. Aging is the main risk factor of neurodegenerative diseases.

Increasing evidence points to oxidative stress and inflammation as common pathological mechanisms underlying neurodegeneration. The Nrf2-ARE pathway is an intrinsic mechanism of defense against oxidative stress. Many evidences demonstrate the protective role of the Nrf2-ARE pathway in neurodegenerative conditions. Nrf2 is a transcription factor that induces the expression of cytoprotective and detoxification genes, thus reduces oxidative stress and neuroinflammation.

Figure 13.1 illustrates oxidative stress as a result of an imbalance between the production of xenobiotic metabolic reactive intermediates as well as endogenous metabolites and the body's defense mechanism.

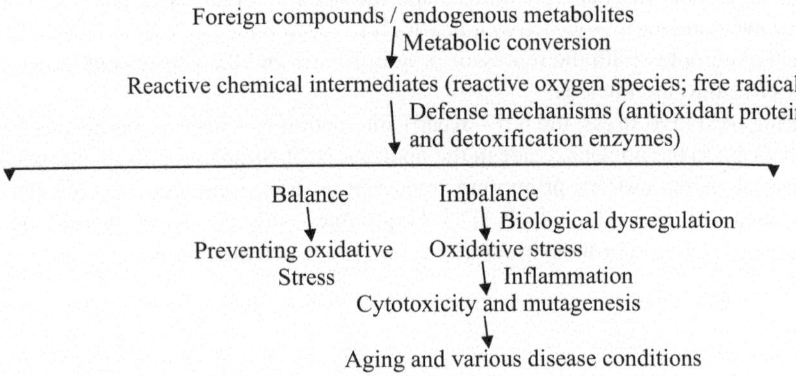

Fig. 13.1 Oxidative stress and biological dysregulation

13.2 Role of Nrf2-ARE Pathway

Nuclear factor erythroid-2-related factor 2 (Nrf2) is a transcription factor that responds to oxidative stress by binding to antioxidant response element (ARE) in the promoter of genes coding for antioxidant proteins and detoxification enzymes. Nrf2-ARE signaling pathway regulates the expression of many genes and functions related to antioxidant proteins and detoxification enzymes. This Nrf2-ARE pathway is critical to the elimination of reactive oxygen species, free radicals, and other reactive chemical intermediates.

As the body's defense mechanism, Nrf2-ARE pathway acts as a sensor to oxidative stress through stimulation of antioxidant proteins and detoxification enzyme genes. Many evidences point to oxidative stress and inflammation as a pathological mechanism underlying neurodegeneration, leading to the protective role of the Nrf2-ARE pathway in neurodegenerative conditions. Targeting the Nrf2-ARE pathway may also present a novel therapeutic approach for the treatments of inflammatory diseases as well as aging-related neurodegeneration and other disease conditions such as cancers.

13.3 Nrf2-Keap1-ARE Pathway in the Absence of Oxidative Stress

Nrf2 is a protein that regulates the expression of antioxidants to protect the body against oxidative stress and its induced inflammation. The primary mechanism of Nrf2 activation is essentially based on the following three steps: (a) the dissociation of Nrf2 from its inactive complex with the repressor protein Keap1, (b) the subsequent translocation of Nrf2 into the nucleus, and (c) the activation of Nrf2 signaling that induces ARE-dependent expression of detoxification enzymes and antioxidant proteins.

Figure 13.2 illustrates Nrf2-Keap1-ARE pathway in the absence of oxidative stress.

Under the non-stressed condition, Nrf2 covalently binds to cysteine residue of repressor Keap1 (Kelch-like ECH associated protein 1) to form the complex Nrf2-Keap1 in the cytoplasm. This Nrf2-Keap1 binding step is followed by the addition of a chain of ubiquitin (Ub, a small regulatory protein) molecules to Nrf2, where Keap1 acts as an essential active protein.

The substrate for ubiquitination is highly specific. Its activity is provided by ubiquitin ligases, where Cullin 3 maintains a staging for ubiquitin ligases. The addition of ubiquitin molecules is referred to as ubiquitination. Ubiquitination affects Nrf2 transcription factor by initialing it for degrading unneeded Nrf2 via proteasomes. Proteasomes are protein complexes which degrade unneeded proteins by proteolysis catalyzed by proteases. In the absence of oxidative stress, Nrf2 is

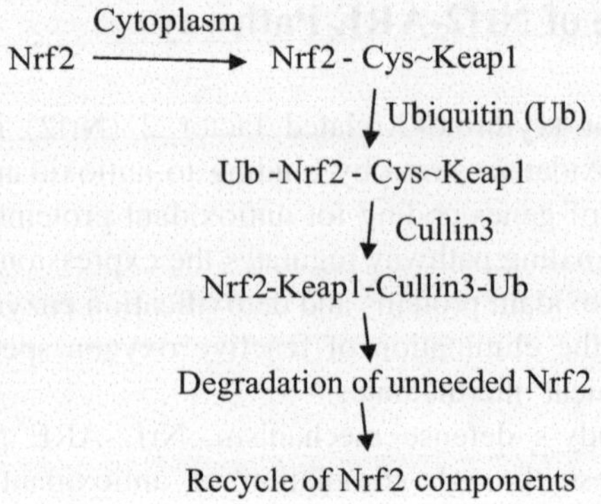

Fig. 13.2 Nrf2-Keap1-ARE pathway in the absence of oxidative stress

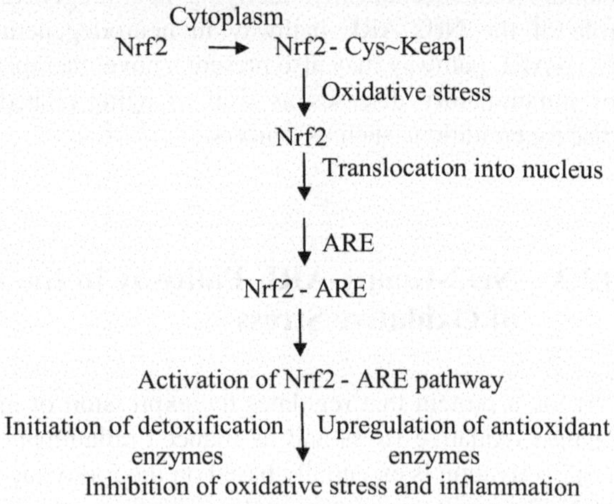

Fig. 13.3 Nrf2-Keap1-ARE pathway in the presence of oxidative stress

constantly degraded through ubiquitination, and the degraded components are recycled in non-stressed cells.

13.4 Nrf2-Keap1-ARE Pathway in the Presence of Oxidative Stress

Nrf2-ARE signaling pathway regulates the expression of genes whose protein products are involved in the detoxification and elimination of reactive intermediates including reactive oxygen species and free radicals produced during bioactivation of foreign compounds and endogenous metabolites. Figure 13.3 illustrates the Nrf2-Keap1-ARE pathway under oxidative stress, where the nuclear translocation of Nrf2 is essential in the defense against oxidative stress.

In the presence of oxidative stress, cysteine residues on Keap1 are modified, resulting in the stabilization and the translocation of Nrf2 into the nucleus. In the nucleus, Nrf2 binds to the promoter region of ARE, which initiates the transcription of cytoprotective detoxification enzymes and upregulates antioxidant function. Following the exposure of Keap1 to oxidative stress, the Keap1-Cullin 3 ubiquitination system is disrupted, and Nrf2 is not degraded. Upon the nuclear entry, Nrf2 binds to ARE in the promoter region of antioxidative genes, which initiates the transcription of antioxidative genes and the corresponding proteins.

One potential mechanism in the defense against oxidative stress is the induction of Nrf2 by small molecule activators to modify Keap1 cysteine residues and to disrupt the Nrf2-Keap1 complex, thus favoring the dissociation of Nrf2 and subsequent nuclear translocation. Another potential mechanism is the involvement of inhibiting Nrf2 ubiquitination by Keap1 and degrading by the proteasome to enhance the availability of Nrf2, thus favoring nuclear translocation of Nrf2.

13.5 Activation of Nrf2-ARE Pathway

The exposures to toxicants mediated with oxidative stress may lead to inflammation and various disease conditions, such as cardiovascular and neurodegenerative disorders and cancers. Nrf2-ARE pathway regulates the transcription of many antioxidant and detoxification genes, so as to process and eliminate toxicants and carcinogens before they can cause damages to the cells. Therefore, to prevent the accumulation of reactive chemical intermediates and reactive oxygen species, Nrf2-ARE pathway must function constantly.

Nrf2 is known for its ability to regulate the expression of a series of metabolic enzymes for antioxidation and detoxification, such as glutathione S-transferase, UDP-glucuronosyltransferase, and NAD(P)H quinone oxidoreductase. Nrf2-ARE activators are potential cytoprotective agents, and the activation of Nrf2-ARE signaling pathway is considered as a major mechanism in the cellular defense against oxidative stress.

Activating antioxidant systems in cells through Nrf2-ARE signaling pathway has been shown to prevent cardiovascular disease by obviating inflammation caused by reactive chemical intermediates and reactive oxygen species, thus leading to relieving oxidative stress and limiting cardiac injury and heart failure. However, over activation of Nrf2-ARE pathway was reported to cause cardiovascular injury. Over activation of Nrf2 in cancer cells has also been implicated in cancer progression and in resistance to cancer chemotherapeutics.

13.6 Cytoprotection and Nrf2-ARE Pathway

To avoid cellular damages caused by cytotoxicity and carcinogenic injury, it is essential to reduce the formation of reactive chemical intermediates such as reactive oxygen species and free radicals. Cytoprotection is considered as a good strategy to reduce oxidative stress. In the liver, Nrf2 regulates cytoprotective genes, which confers cellular resistance to toxicity and oxidative stress, thus providing protection against inflammation and various disease conditions. Induction of Nrf2-ARE-mediated expressions of antioxidants and detoxification enzymes with cytoprotective agents may counteract the toxic effects of chemically active species.

13.6.1 *Induction of Nrf2-ARE Pathway*

Metabolic activation and detoxification play a critical role in bioactivation, detoxication, and excretion of foreign compounds and endogenous metabolites from the body. In many cases, activation reactions during biotransformation result in the generation of reactive intermediates, including reactive oxygen species and free radicals, and many of them may result in cytotoxicity and carcinogenicity. To counteract such harmful effects, living cells evolve sophisticated signaling mechanisms in response to oxidative stress mediated by such reactive chemical species.

Detoxification enzymes have long been investigated for potentially preventative and protective effects against oxidative stress-related cytotoxicity and carcinogenicity. As a result, induction and activation of antioxidants and detoxification enzymes through Nrf2-ARE signaling pathway represents a critical aspect of cytoprotective mechanism of action.

13.6.2 *Detoxification Enzyme Inducers for Cytoprotection*

Detoxification enzyme inducers such as phytochemicals are compounds mainly produced by plants and widely distributed in fruits and vegetables. A strategy for chemoprevention is focused on the use of inducers to modulate the metabolism of foreign compounds such as drugs and other chemicals through upregulation of metabolic detoxification enzymes. Figure 13.4 illustrates the inducers of detoxification enzymes as a strategy for cytoprotection of cells against oxidative damages.

Fig. 13.4 Induction of Nrf2-ARE Pathway for Cytoprotection

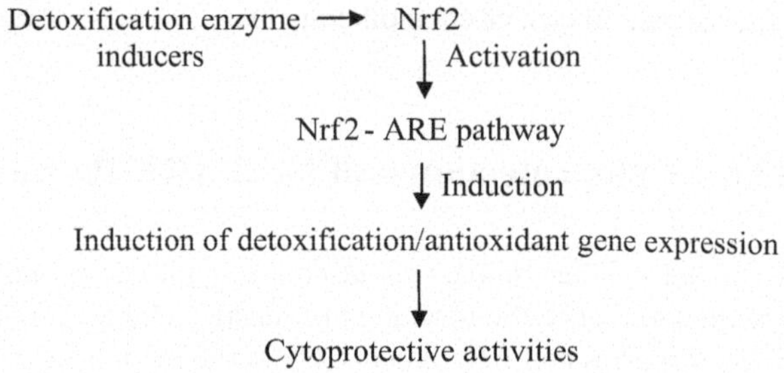

Nrf2-ARE pathway is an interesting target for inducers of metabolic activation enzymes. Understanding the relationship between Nrf2-ARE and metabolism in cellular tissues may provide novel insights for interventions in protection against cytotoxicity and carcinogenesis. Research has revealed that dietary phytochemicals possess potential to protect organisms against cellular damages caused by oxidative stress, due to their capacities to induce antioxidants and detoxification enzymes.

13.6.3 Dietary Phytochemicals

Extensive studies revealed that Nrf2- RE can be induced by a variety of small molecules. Accordingly, there has been a growing interest in investigating novel Nrf2-ARE activators from dietary sources. Many of such inducers are present in human daily diets. Diet-derived compounds studied for cytoprotective activities are resveratrol from grapes and broccoli, curcumins from turmeric, flavonoid genistein from soy, and quercetin from onion and green tea. Other phytochemicals include polyunsaturated fatty acids, carotenoids such as β-carotene, lycopene and lutein, indole-3-carbinol, isothiocyanates, and pomegranate.

Many inducers with cytoprotective properties exhibit antioxidant and anti-inflammatory functions through activation of Nrf2-ARE pathway, which induces detoxification enzyme and antioxidant genes. Sulforaphane, an isothiocyanate derived from cruciferous vegetables, is the most extensively studied phytochemical in the literature. Sulforaphane reacts with protein thiols to form thionoacyl adducts, thus affecting the Cys residues in Keap1 protein and modulating Nrf2 protein stability. Isothiocyanate from Japanese horseradish was also reported to activate ARE, mediate Nrf2 binding to ARE, and induce detoxification enzyme genes.

The induction of detoxification enzyme genes was also mediated by major garlic components such as diallyl sulfide, disulfide, and trisulfide. Moreover, 1,2-dithiole-3-thione, the simplest member of sulfur-containing dithiolethiones found in cruciferous vegetables, was also reported to trigger nuclear accumulation of the transcription factor Nrf2 and enhance its binding to ARE. Nrf2 binding to ARE activates detoxification enzyme genes involved in detoxifying toxicants and carcinogens, leading to the attenuation of oxidative stress.

An increase in NAD(P)H quinone dehydrogenase 1 transcription in response to phytochemical quercetin suggests that dietary plant polyphenols can stimulate the transcription of detoxification enzymes, thus offers a protection against carcinogenic chemicals. Moreover, dietary curcumin was found to induce the activities of glutathione S-transferase and NADH:quinone oxidoreductase-1(NQO1), leading to increased detoxification of toxic benzo[a]pyrene. Quercetin and phenylpropanoids in ginger are also known chemoprotective compounds that can significantly increase the level of Nrf2 activity.

ARE in accompany with transcription factor Nrf2 is also involved in the modulation of NQO1 gene expression by resveratrol. Moreover, anthocyanins among phytochemical flavonoids were reported to stimulate the antioxidant defense system, leading to the protection against oxidant-induced injury through affecting the expression of ARE-regulated detoxification enzymes and the modulation of antioxidative stress defense system.

Furthermore, dibenzoylmethane, a minor constituent of licorice, is a beta-diketone phytochemical with a wide variety of anticancer effects. Dibenzoylmethane was found to increase the DNA binding activity of Nrf2, mediate the induction of detoxification enzymes by Nrf2 activation, and inhibit the formation of benzo[a]pyrene-induced DNA adducts. In addition, licorice from the root of plant *Glycyrrhiza uralensis* has been used as traditional medicine to soothe gastrointestinal problems and to repair stomach lining through its anti-inflammatory and immune-boosting properties.

Other phytochemicals include zerumbone (ZER), a sesquiterpene compound occurring in tropical ginger. ZER has been implicated as a promising chemoprotective agent against colon and skin cancer. It was found to induce the transcription factor Nrf2 that binds to antioxidant response element (ARE). This finding suggested that ZER is a potential activator of the Nrf2-ARE-dependent detoxification pathway.

A more detailed discussion of dietary phytochemicals in vegetables and fruits is further discussed in following chapters.

13.7 Over Activation of Nrf2-ARE Pathway

As discussed above, the Keap1-Nrf2-ARE pathway is one of the most important defense mechanisms against oxidative stress. This pathway is closely associated with inflammatory diseases. Most antioxidants activate the Nrf2-ARE antioxidant pathway, thus helping prevent inflammation and disease conditions caused by oxidative stress. Activation of Nrf2 signaling induces ARE-dependent expression of detoxification and antioxidant defense proteins. Nrf2-ARE signaling has become an attractive target for cancer chemoprevention. It needs to note that, when overactivated, the Nrf2-ARE antioxidant system can cause various disease conditions, including autoimmune, neurodegeneration, cardiovascular diseases, and aging. Moreover, constitutive over activation of Nrf2 in cancer cells has been implicated in cancer progression as well as in resistance to cancer chemotherapeutics.

Bibliography

Bayele HK, Debnam ES, Srai KS (2016) Nrf2 transcriptional derepression from Keap1 by dietary polyphenols. Biochem Biophys Res Commun 469:521–528

Bianco NR, Chaplin LJ, Montano MM (2005) Differential induction of quinone reductase by phytoestrogens and protection against oestrogen-induced DNA damage. Biochem J 385:279–287

Bryan HK, Olayanju A, Goldring CE, Park BK (2013) The Nrf2 cell defence pathway: Keap1-dependent and -independent mechanisms of regulation. Biochem Pharmacol 85(6):705–717

Buendia I, Michalska P, Navarro E, Gameiro I, Egea J, León R (2016) Nrf2-ARE pathway: an emerging target against oxidative stress and neuroinflammation in neurodegenerative diseases. Pharmacol Ther 157:84–104

Chen CH (2012) Activation and detoxification enzymes: functions and implications. Springer Sciences, New York

Copple IM, Goldring CE, Kitteringham NR, Park BK (2010) The keap1-nrf2 cellular defense pathway: mechanisms of regulation and role in protection against drug-induced toxicity. Handb Exp Pharmacol 196:233–266

de Vries HE, Witte M, Hondius D, Rozemuller AJ, Drukarch B, Hoozemans J, van Horssen J (2008) Nrf2-induced antioxidant protection: a promising target to counteract ROS-mediated damage in neurodegenerative disease? Free Radic Biol Med 45(10):1375–1383

Johnson JA, Johnson DA, Kraft AD, Calkins MJ, Jakel RJ, Vargas MR, Chen PC (2008) The Nrf2-ARE pathway: an indicator and modulator of oxidative stress in neurodegeneration. Ann N Y Acad Sci 1147:61–69

Joshi G, Johnson JA (2012) The Nrf2-ARE pathway: a valuable therapeutic target for the treatment of neurodegenerative diseases. Recent Pat CNS Drug Discov 7(3):218–229

Kaspar JW, Niture SK, Jaiswal AK (2009) Nrf2:INrf2(Keap1) signaling in oxidative stress. Free Radic Biol Med 47(9):1304–1309

Kensler TW, Curphey TJ, Maxiutenko Y, Roebuck BD (2000) Chemoprotection by organosulfur inducers of phase 2 enzymes: dithiolethiones and dithiins. Drug Metabol Drug Interact 17:3–22

Keum YS (2011) Regulation of the Keap1/Nrf2 system by chemopreventive sulforaphane: implications of posttranslational modifications. Ann N Y Acad Sci 1229:184–189

Krajka-Kuźniak V, Paluszczak J, Baer-Dubowska W (2017) The Nrf2-ARE signaling pathway: an update on its regulation and possible role in cancer prevention and treatment. Pharmacol Rep 69 (3):393–402

Kwak MK, Egner PA, Dolan PM, Ramos-Gomez M, Groopman JD, Itoh K, Yamamoto M, Kensler TW (2001) Role of phase 2 enzyme induction in chemoprotection by dithiolethiones. Mutat Res 480–481:305–315

Lee JM, Johnson JA (2004) An important role of Nrf2-ARE pathway in the cellular defense mechanism. J Biochem Mol Biol 37:139–143

Lee JS, Surh YJ (2005) Nrf2 as a novel molecular target for chemoprevention. Cancer Lett 224 (2):171–184

Li J, Calkins MJ, Johnson DA, Johnson JA (2007) Role of Nrf2-dependent ARE-driven antioxidant pathway in neuroprotection. Methods Mol Biol 399:67–78

Li W, Guo Y, Zhang C, Wu R, Yang AY, Gaspar J, Kong AN (2016) Dietary phytochemicals and cancer chemoprevention: a perspective on oxidative stress, inflammation, and epigenetics. Chem Res Toxicol 29:2071–2095

Lu MC, Ji JA, Jiang ZY, You QD (2016) The keap1-nrf2-ARE pathway as a potential preventive and therapeutic target: an update. Med Res Rev 36(5):924–963

Morimitsu Y, Nakagawa Y, Hayashi K, Fujii H, Kumagai T, Nakamura Y, Osawa T, Horio F, Itoh K, Iida K, Yamamoto M, Uchida K (2002) A sulforaphane analogue that potently activates the Nrf2- dependent detoxification pathway. J Biol Chem 277:3456–3463

Nakamura Y, Yoshida C, Murakami A, Ohigashi H, Osawa T, Uchida K (2004) Zerumbone, a tropical ginger sesquiterpene, activates phase II drug metabolizing enzymes. FEBS Lett 572:245–250

Nguyen T, Nioi P, Pickett CB (2009) The Nrf2-antioxidant response element signaling pathway and its activation by oxidative stress. J Biol Chem 284(20):13291–13295

Nishinaka T, Ichijo Y, Ito M, Kimura M, Katsuyama M, Iwata K, Miura T, Terada T, Yabe-Nishimura C (2007) Curcumin activates human glutathione S-transferase P1 expression through antioxidant response. Toxicol Lett 170:238–247

Qin S, Hou DX (2016) Multiple regulations of Keap1/Nrf2 system by dietary phytochemicals. Mol Nutr Food Res 60:1731–1755

Saw CL, Kong TA (2011) Nuclear factor-erythroid 2-related factor 2 as a chemopreventive target in Colorectal cancer. Expert Opin Ther Targets 15(3):281–295

Schadich E, Hlavac J, Volna T, Varanasi L, Hajduch M, Dzubak P (2016) Effects of ginger phenylpropanoids and quercetin on Nrf2-ARE pathway in human BJ fibroblasts and HaCaT keratinocytes. Biomed Res Int 2016:2173275

Shih PH, Yeh CT, Yen GC (2007) Anthocyanins induce the activation of phase II enzymes through the antioxidant response element pathway against oxidative stress-induced apoptosis. J Agric Food Chem 55:9427–9435

Smith RE (2019) The effects of dietary supplements that overactivate the Nrf2/ARE system. Curr Med Chem

Smith RE, Tran K Smith CC, McDonald M, Shejwalkar P, Hara K (2016) The role of the Nrf2/ARE antioxidant system in preventing cardiovascular diseases. Diseases 4(4):34

Stefanson AL, Bakovic M (2014) Dietary regulation of Keap1/Nrf2/ARE pathway: focus on plant-derived compounds and trace minerals. Nutrients 6:3777–3801

Sun Y, Yang T, Leak RK, Chen JH, Zhang F (2017) Preventive and protective roles of dietary Nrf2 activators against central nervous system diseases. CNS Neurol Disord Drug Targets 16 (3):326–338

Tan XL, Spivack SD (2009) Dietary chemoprevention strategies for induction of phase II xenobiotic- metabolizing enzymes in lung carcinogenesis: a review. Lung Cancer 65:129–137

Thimmulappa RK, Rangasamy T, Alam J, Biswal S (2008) Dibenzoylmethane activates Nrf2-dependent detoxification pathway and inhibits benzo(a)pyrene induced DNA adducts in lungs. Med Chem 4(473):81

Chapter 14
Genetic Variations and Polymorphisms of Metabolic Enzymes

Genetic polymorphisms are one of two important features of metabolic enzymes. Another important feature is the inducibility of metabolic enzymes. Genetic polymorphisms affect not only the metabolism of foreign compounds but also the responses of metabolic reactive intermediates to cellular damages. Enzyme polymorphisms therefore play a critical role in xenobiotic cytotoxicity and carcinogenesis as well as xenobiotics-modulated disease conditions. The subject of genetic polymorphisms of metabolic enzymes is discussed below, while the inducibility of metabolic enzymes is presented in the following chapters.

DNA sequence refers to the precise order of nucleotide bases that encode amino acids. Genetic polymorphism is a DNA sequence variation, which is common in human population. The genetic code relates the sequence of bases in DNA to the sequence of amino acids in proteins. The gene that encodes a specific protein may have a difference in DNA sequences.

Genes involved in the biosynthesis of metabolic enzymes are subject to genetic polymorphisms.

Genetic polymorphisms give rise to differences in the ability of individuals to the effectiveness of metabolism of foreign compounds. The metabolism of drugs and other foreign compounds into more hydrophilic metabolites is essential to facilitate their excretion from the body. Genetic differences in drug metabolism are the result of variations in alleles for genes that code for the enzymes responsible for drug metabolism.

Genetically determined differences in the activity of the metabolic enzymes can increase an individual's susceptibility to certain types of chemically induced diseases. Interindividual differences in genes for drug metabolizing enzymes are important for understanding efficacy in drug therapy. These differences are also important for the detection of individuals who are prone to side effects from the drug.

Enzymes are frequently polymorphic with two or more variants of an enzyme encoded by a single locus. The variants differ slightly in their amino acid sequence. Hence, genetic polymorphism may act as molecular biomarkers to provide important predictive information about an individual's response to cytotoxicity and

C.-H. Chen, *Xenobiotic Metabolic Enzymes: Bioactivation and Antioxidant Defense*,
https://doi.org/10.1007/978-3-030-41679-9_14

carcinogenesis. Investigations of the effects of enzyme polymorphisms on chromosome damage as biomarkers of genotoxic exposure may facilitate the identification of disease risk groups.

14.1 Enzyme Polymorphisms

Enzyme polymorphism may contain two or more variants of an enzyme encoded by a single locus. Although the variant differs slightly in the amino acid sequence, it may affect substrate specificity, enzymatic reaction activity, as well as enzymatic-related transport functions. Owing to the altered activity of the encoded enzyme, polymorphisms in genes coding to metabolic enzymes can result in variation of activation and detoxification efficiency of foreign compounds. As a result, genetic polymorphisms may lead to altering individual's response to cytotoxicity and carcinogenesis of foreign compounds.

Altered expression or activity of the encoded proteins implies that enzyme polymorphisms can influence how humans metabolize foreign compounds including foods, drugs, and other chemicals. Reduction in metabolizing capacity of activation enzymes can lead to lower metabolic rate, and reduction of metabolic capacity of detoxification enzymes can result in increased toxic effects of reactive intermediate species derived from xenobiotic metabolism.

Individual predisposing factors due to genetic polymorphisms in detoxification enzymes have been implicated in the molecular pathogenesis and the heterogeneity of many diseases. Dietary or environmental carcinogens can bind to DNA, resulting in DNA mutation and leading to carcinogenesis. As a result, the interrelationship between genetic susceptibility and carcinogenic exposure plays an important role in cancer development.

14.2 Genetic Polymorphisms of Xenobiotic Metabolic Enzymes

A wide range of xenobiotic-metabolizing enzymes are responsible for the metabolism of foreign compounds digested through diets and drugs or inhaled from the environment. Metabolic enzyme-catalyzed reactions serve as activation and detoxification processes to make foreign compounds and endogenous metabolites more soluble in water and less toxic, thus facilitating their excretion from the body. The polymorphic expression of metabolic enzymes attributes in part to interindividual variability in the metabolism of toxicants and carcinogens.

Biotransformation plays an important role in the activity and the specificity of carcinogens. Figure 14.1 illustrates the effects of genetic polymorphisms on metabolic activation enzymes. A higher activation capacity can significantly increase

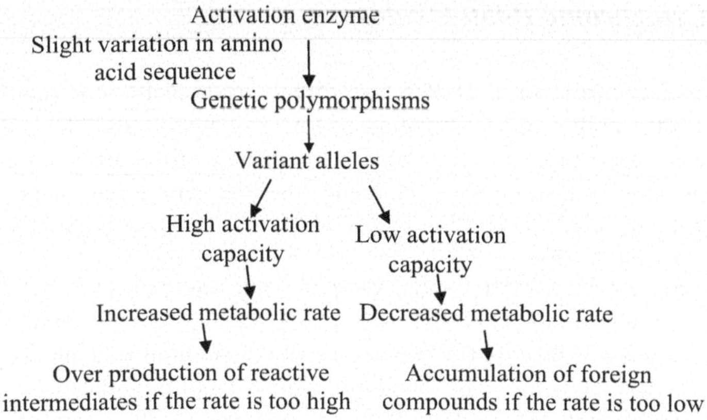

Fig. 14.1 Activation enzyme genetic polymorphisms affecting bioactivation of foreign compounds

Fig. 14.2 Detoxification enzyme genetic polymorphisms affecting foreign compound-mediated toxicity

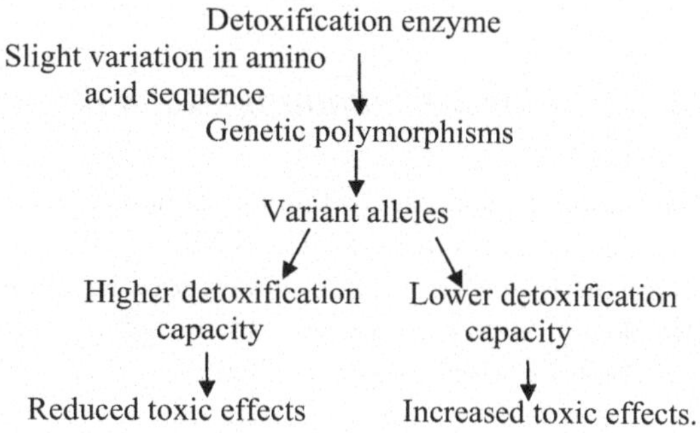

metabolic rate. But, a much higher activation rate may lead to overproduction of reactive intermediates or metabolites.

An allele is one of the forms of gene that appears at a particular location on a specific chromosome. Figure 14.2 presents the effects of genetic polymorphisms on metabolic detoxification enzymes. A variant allele with a higher detoxifying capacity can lead to reduced toxic effects, while that with a lower detoxifying capacity can result in increased toxic effects.

14.3 Polymorphisms of Activation Enzymes

Variants in the gene encoding of activation enzymes can result in altering enzymatic activity and bioactivation of foreign compounds. Genetic polymorphisms of major activation enzymes, such as cytochrome P450 oxidase, flavin-containing monooxygenase, peroxidase, and carboxylesterase, and their associated variations in activation metabolism are discussed below.

14.3.1 Cytochrome P450 Oxidase

Genetic polymorphisms of CYP450s are largely responsible for interindividual differences in the ability of bioactivation of foreign compounds. The polymorphic responsiveness appears to be correlated with genetic differences in the risk of bronchogenic carcinoma caused by cigarette smoking. The tobacco-specific nitrosamines require metabolic activation by CYP450s to generate reactive intermediates that induce DNA mutation.

The potential role of CYP2E1 and CYP2D6 in nitrosamine-induced damages has been characterized, where CYP2E1 is CYP450 of family 2, subfamily E, and polypeptide 1 and CYP2D6 is CYP450 of family 2, subfamily D, and polypeptide 6. CYP2E1 polymorphism was found to link to smoking-related lung cancer. Moreover, a highly induced level of CYP1A1 genes can result in greater concentrations of polycyclic aromatic reactive intermediates that form DNA adducts.

14.3.2 Flavin-Containing Monooxygenase

The majority of flavin-containing monooxygenase (FMO)-mediated metabolism in the liver is catalyzed by FMO3 (FMO family 3 polymorphism). FMO3 is a hepatic microsomal enzyme that oxidizes foreign compounds such as drugs and other chemicals. Numerous variants in the gene encoding FMO3 have been identified and some of which result in altered enzymatic activity and substrate metabolism. Two commonly occurring polymorphisms of FMO3 are E158K (glutamic acid is replaced with lysine at position158) and E308G (glutamic acid is replaced with glycine at position 308), which have been found to associate with a reduction in polyp burden in patients with adenomatous polyposis.

14.3.3 Peroxidase

Myeloperoxidase (MPO), an enzyme highly expressed in leukocytes, transforms procarcinogens, such as benzo[a]pyrene and aromatic amines, to reactive intermediates. The MPO463A (alanine at position 463) polymorphism in the promoter region strongly reduces MPO mRNA expression, suggesting that MPO463A variant may be a protective factor in the etiology of lung cancer.

14.3.4 Carboxylesterase

Carboxylesterasc1 (CES1; CES family 1) and carboxylesterase2 (CES2; family 2) polymorphisms are responsible for the hydrolysis of ester and amide bonds present in pharmaceutical products. Genetic variants of CES1 and CES2 genes have been shown to influence drug metabolism. Polymorphisms of the CES1 gene have been reported to affect the metabolism of methylphenidate, a stimulant medication, whereas variants of the CES2 gene have been found to affect the metabolism of aspirin (Table 14.1).

14.4 Polymorphisms of Detoxification Enzymes: Transferases

Detoxification enzymes are mainly transferases, including glutathione S-transferases, UDP-glucuronosyltransferases, sulfotransferases, N-acetyltransferases, and methyltransferases. Table 14.2 lists genetic polymorphisms of detoxification enzymes, which association with foreign compound-mediated cytotoxicity and carcinogenesis is discussed below.

Table 14.1 Typical genetic polymorphisms of activation enzymes

Activation enzymes	Genotype
Cytochrome P450 oxidase	CYP2E1, CYP2D6, CYP1A1
Flavin-containing monooxygenase	FMO3 E158K, FMO3 E308G
Peroxidase	MPO463A
Carboxylesterase	CES1, CES2

Table 14.2 Typical genetic polymorphisms of detoxification enzymes

Detoxification enzymes	Genotype
Transferases	
Glutathione S-transferase	GSTM1, GSTM1 mull
UDP-glucuronosyltransferase	UGT1A, UGT2A, UGT2B
Sulfotransferase	SULT1A1*1, SULT1A1*2
N-acetyltransferase	NAT1, NAT2
Methyltransferase	COMTVal158Met
Non-transferases	
Quinone oxidoreductase	NQO1C609T
Epoxide hydrolase	EPHXT113C, EPHXA139G

14.4.1 Glutathione S-Transferase

Glutathione S-transferase (GST) genes encode a family of detoxification enzymes that are involved in the detoxification metabolism by conjugating foreign compound with glutathione. GST-catalyzed biotransformation reactions facilitate the excretion of drugs and other chemicals, thus offering protection against toxicants derived from exogenous reactive chemical intermediates and endogenous metabolites.

Genetic polymorphisms in GST enzymes may result in variations in their detoxification activities. Variations in GST genes can alter the catalytic efficiency of GST isoenzymes. Two relevant GST polymorphisms, the genotypic frequencies of GST theta1 (GSTT1) and GST mu1 (GSTM1), play a role in the detoxifying reactive oxygen species that are involved in cellular processes of inflammation, aging, and degenerative diseases. It was reported that the homozygous deletions or null genotypes of GSTT1 and GSTM1 genes may be associated with an increased risk of cancer.

GSTM1 deletion polymorphisms (GSTM1 null genotype) lead to an increased oxidative stress compared with wild genotype, resulting in DNA damage significantly higher in GSTM1 null genotype than wild genotype. GSTM1 null genotype was found to increase DNA adduct levels in the lung, leading to an increased lung cancer risk. GSTM1 null genotype may also predispose a modest increase in the risk of bladder cancer. In contrast, no similar association was observed with polymorphisms of other GST genes such as GSTT1 and GSTP1.

A higher level of DNA adducts and chromosome damage has been detected in lymphocytes of tobacco smokers and bus drivers who are lacking of the GSTM1 gene. In addition, GSTM1 null genotype was also reported to significantly associate with increased susceptibility of developing gastric cancer in Asians. The prevalence of GSTM1 null genotype was found in 48% of gastric cancer cases. Furthermore, a meaningful increase in breast cancer risk in association with GSTM1 null genotype has also been reported.

The GSTM1 homozygous null genotype also exhibits an increased risk of hypertension, but not for GSTT1 null genotype. Nevertheless, GSTT1 deletion was found to relate to genetic predisposition to lymphoma, particularly interacting with environmental pollutants containing polycyclic aromatic hydrocarbons. There are also evidences suggesting that detoxification enzymes may play a role in prostate cancer development. Investigations of genetic polymorphisms in GSTM1 and GSTT1 have revealed a significantly higher risk of prostate cancer with null allele of GSTT1 and GSTM1.

14.4.2 UDP-Glucuronosyltransferase

UDP-glucuronosyltransferases (UGTs) are a major metabolizing enzyme involving in the detoxification of a large number of drugs, environmental chemicals, and endogenous compounds. UGTs are expressed predominantly in the liver and gastrointestinal tract. UGTs contribute to the detoxification of drugs by forming water-soluble glucopyranosiduronic acids. Glucuronidation reactions catalyzed by UGT enzymes attribute to a detoxification process that adds glucuronic acid to exogenous and endogenous compounds, thus reducing the extent of their toxicity and facilitating their excretion from the body.

Glucuronidation is a crucial pathway of metabolism that detoxifies reactive chemical intermediates including reactive oxygen species. Due to genetic polymorphisms and differences in the regulation of UGT genes, the expression of UGT varies widely between individuals. UGT function is genetically modulated by nucleotide polymorphisms which lead to the expression of functionally altered enzymes. Genetic polymorphisms in UGT may result in variations in its detoxification activities, which subsequently might influence the levels of toxicants and carcinogenic compounds.

It has been hypothesized that UGT polymorphisms that reduce the capacity to glucuronidate carcinogens are associated with an increased risk of developing cancers.

Identification of genetic polymorphisms that attribute to the risk of developing cancers is important for cancer prevention. Modulation of UGT enzymatic activity or gene promoter activity has been implicated as a risk factor for adverse drug effects and for the development of cancer.

UGT consists of two subfamilies, UGT1 (UGT, family 1) and UGT2 (UGT, family 2). One of the subfamilies of UGT1 enzyme is UGT1A, which is a key hepatic detoxification enzyme involving biotransformation of many carcinogens. The functional UGT1A isoforms are encoded by a single UGT1A gene locus with multiple exons. UGT1A1 (UGT, family 1, subfamily A, polypeptide 1) plays a significant role in the detoxification of estrogens and may serve as a therapeutic target for the prevention and treatment of breast cancer and other estrogen-related diseases.

Catechol estrogen metabolites can induce carcinogenesis by acting as endogenous tumor initiators. Glucuronidation, mediated by UGT1A1 enzyme, is a main metabolic pathway of estrogen detoxification in steroid target tissues. Among different variants of UGT1A1 is UGT1A1*6 (UGT, family 1, subfamily A, polypeptide 1, allele 6) polymorphism, which has been deduced to be a risk factor for breast cancer in women of Chinese ethnicity.

Individual susceptibility to colorectal cancer may be in part due to variations in detoxification capacity in the gastrointestinal tract. Variant UGT1A6 and UGT1A7 genotypes with reduced enzymatic activities might enhance susceptibility to colorectal cancer. Genetic polymorphisms in UGT1A7 are vulnerable to environmental carcinogenesis and are implicated as a cancer risk gene. UGT1A7*3 allele encoding a low catalytic activity may act as a risk gene in proximal digestive tract cancer and is considered a potential marker for colorectal cancer risk.

UGT-catalyzed glucuronidation is a primary detoxification pathway of dietary heterocyclic amines and major tobacco carcinogens, polycyclic aromatic hydrocarbons. Genetic polymorphisms in genes that encode enzymes involved in metabolizing tobacco carcinogens could affect an individual's risk for lung cancer. UGT1A6 may be used to identify people with increased risk of developing lung cancer. Moreover, UGT1A7, a detoxifying enzyme capable of tobacco-borne carcinogen detoxification and cellular protection, has been implicated as a cancer risk gene. UGT1A9 has been shown to play an important role in the clearance of anticancer and pain medications, while UGT1A10 exhibits significant glucuronidation activity against the metabolites of polycyclic aromatic hydrocarbons and benzo[a]pyrene.

UGT2 family is subdivided into two subfamilies, UGT2A and UGT2B. Polymorphisms of UGT2A and UGT2B genes have been shown to modulate UGT enzymatic activity. UGT2A is an important detoxification enzyme in the metabolism of polycyclic aromatic hydrocarbons from tobacco carcinogens. Functional polymorphisms in UGT2A and UGT2B may play a role in tobacco-related cancer risk. Moreover, UGT2B10 and UGT2B17 polymorphisms are associated with a reduced detoxification capacity against tobacco-specific nitrosamine and may therefore affect individual cancer risk upon exposure to tobacco. The association of the UGT2B17 deletion with increased lung adenocarcinoma in women is consistent with its association with decreased glucuronidation rates.

Furthermore, a significant association was found between the UGT2B17 deletion genotype and decreased colorectal cancer risk. Genetic polymorphism in UGT2B7 may predispose to colorectal cancer. Moreover, polymorphisms in the UGT genes may alter enzyme activity and subsequently play a role in esophageal cancer etiology. Environmental and dietary factors also play crucial roles in esophageal carcinogenesis. UGT1A8 and UGT2B4 genotypes with decreased enzyme activities were reported to be significantly associated with an increased risk of esophageal cancer.

14.4.3 *Sulfotrasferase*

Sulfotransferase (SULT) catalyzes sulfate conjugation of various phenolic derivatives of foreign compounds including drugs and other chemicals as well as endogenous substances such as neurotransmitters and steroid hormones. Genetic polymorphisms in three alleles of human SULT1A1 gene are SULT1A1∗1, SULT1A1∗2, and SULT1A1∗3. Among them, SULT1A1∗2 is a polymorphism where the amino acid at codon 213 changes from Arg to His. This His allele (SULT1A1∗2) encodes the enzyme with much lower catalytic activity than the enzyme encoded with the Arg allele (SULT1A1∗1). Consequently, SULT1A1∗2 has an important functional impact on diverse sulfate substrates.

Marked differences in the activities of SULT1A1 variants toward a variety of substrates are critical for understanding interindividual variability of SULT in response to toxicants and carcinogens. The low activity SULT1A1∗2 allele may be at higher risk following carcinogen exposure than that with the SULT1A1∗1 genotype. Individuals with the lower activity SULT1A1∗2 allele may be at a higher cancer risk following carcinogen exposure than those with the higher activity SULT1A1*1 genotype. Moreover, SULT1A1*2 polymorphism enhances the effect of endogenous estrogen exposure, suggesting that this allele polymorphism may be a risk factor for breast cancer.

SULT1A1 is involved in the bioactivation of N-hydroxy metabolite of dietary procarcinogen heterocyclic amines. There were reports that the variant allele of the SULT1A1 gene may be associated with lung cancer, particularly among those who have ever smoked. Comparative studies of SULT1A1 gene polymorphism reveal that variations in SULT1A1 activity may also attribute to prostate cancer risk.

14.4.4 N-Acetyltransferase

N-Acetyltransferase (NAT)-catalyzed acetylation reactions play an essential role in the detoxification of aromatic and heterocyclic amines in foreign compounds, including therapeutic drugs and environmental chemicals. The acetylation polymorphism is a common inherited variation in human drug and carcinogen metabolism. The genes that encode NAT1 and NAT2 have a variety of high and low activity alleles. NAT1 and NAT2 polymorphisms have been shown to be a potential source of adverse drug reactions. Genetic polymorphisms in NATs may also act as molecular biomarkers to provide important predictive information about cytotoxicity and carcinogenesis.

Aromatic amines and their metabolites are a class of carcinogens, especially present in tobacco smoke. Amines undergo metabolism through N- or O-acetylation pathway catalyzed by NAT1 or NAT2. NAT1 and NAT2 are important for detoxification of carcinogenic aromatic and heterocyclic amines. These enzymes also play a key role in metabolism of carcinogenic amines present in overcooked red meat. Moreover, NAT2 gene variants associated with a slow acetylation phenotype are more susceptible to the effects of tobacco smoking with respect to adenoma risk.

Polymorphisms in NAT2 have been suggested as susceptibility factors for DNA damage and lung cancer. Lung adduct levels are significantly increased for the NAT2 slow acetylation genotype, which was observed more frequently in lung cancer patients. The combination of NAT2 slow genotype with glutathione S-transferases GSTM1 null genotype noticeably enhances the levels of lung adducts, suggesting a synergistic effect on DNA damage in the lung tissue.

Moreover, exposures to carcinogens present in diets, cigarette smoke, or environment chemicals have been associated with increased risk of bladder cancer. Polymorphisms in NAT2 have been recognized to link between decreased acetylation phenotype and increased risk for bladder cancer. Slow NAT activity generally exhibits a higher adduct level of aromatic amines in bladder tissues, leading to more susceptible to developing bladder cancer. Heavy smokers with NAT2 slow acetylation genotypes were found to show an over sixfold increase in bladder cancer risk.

Colorectal adenoma is a colorectal cancer precursor. Many studies have focused on genetic variation in NAT2 and its potential as a risk factor in colorectal cancer. Moreover, women who smoke and have a genetically NAT2 slow acetylation genotype or GSTM1 null genotype are at increased risk of breast cancer. The NAT2 genotype with a slow acetylation genotype was also found to associate with an increase in the risk of cervical cancer.

NAT2 may also have a major role in asthma pathogenesis. Studies of NAT polymorphisms revealed a significant association of slow acetylation phenotype with bronchial and occupational asthma. Such genetic variations may also have important implications in the efficacy of drugs for asthma treatment.

14.4.5 Methyltransferase

Catechol-O-methyltransferase (COMT) catalyzes the transfer of a methyl group from S-adenosylmethionine to a catechol-containing substrate molecule. COMT methylation is a metabolic inactivation pathway for catechol estrogens. Intensive studies have revealed a role for the oxidative metabolites of estrogen, particularly catechol estrogens, in the development of estrogen carcinogenesis.

COMT is polymorphic in human populations. COMT gene polymorphisms have been shown to affect COMT enzyme functions. The variant form of COMT can affect the detoxification efficiency of endogenous or exogenous catechol estrogens. In the COMT Val158Met polymorphism, valine is substituted with methionine at nucleotide position 158. Val158Met may change the secondary or tertiary structure of COMT, making it more susceptible to undergo inhibition and inactivation. The variant Val158Met in the COMT gene was also found to exhibit a several-fold decrease in enzymatic activity.

A low activity of COMT variant can give rise to the accumulation of potentially carcinogenic catechol estrogens and their reactive intermediates, thus causing an increase in the risk of tumorigenesis. It has been reported that a low activity variant of the COMT enzyme is associated with an increased risk of developing breast cancer.

If not inactivated by COMT, catechol estrogens can generate a significant quantity of reactive oxygen species, causing oxidative stress and damages to health. In addition, SULT and UGT genes are also involved in catechol estrogen detoxification through conjugation reactions. A reduction in metabolizing capacity of detoxification enzymes can lead to increased toxic effects from oxidative metabolites of estrogen.

Estrogen has been suggested to trigger breast cancer development via an initiating mechanism involving its metabolite. These findings supported a hypothesis that breast cancer can be initiated by estrogen exposure and that increased estrogen exposure confers a higher risk of breast cancer. Besides the gene, environment interaction is also crucial in modulating the physiological role of the COMT enzyme.

14.5 Polymorphisms of Detoxification Enzymes: Non-transferases

Non-transferase detoxification enzymes include quinone oxidoreductase and epoxide hydrolase. Table 14.2 also lists genetic polymorphisms of these non-transferase enzymes. Their association with foreign compound-mediated cytotoxicity and carcinogenesis is discussed below.

14.5.1 Quinone Oxidoreductase

NADH:quinone oxidoreductase 1 (NQO1) is a flavoprotein that catalyzes detoxification reduction by converting quinoid compounds to hydroquinones, thus protecting cells from cytotoxicity, carcinogenicity, and oxidative stress. NQO1 enzyme also detoxifies environmental carcinogens such as nitroaromatic compounds and heterocyclic amines. Genetic polymorphisms of NQO1 may result in variations in its detoxification activities, leading to influencing the levels of toxicant and carcinogenic compounds present in the body.

A genetic variant in NQO1 may modify individual susceptibility to hepatocellular carcinoma. The gene coding for NQO1 has a genetic polymorphism C609T in which cytosine is substituted with thymine at nucleotide position 609 of the NQO1 cDNA. NQO1 C609T polymorphism reduces NQO1 enzyme activity, thereby diminishing the detoxifying capacity provided by the enzyme. The reduced enzymatic activity in NQO1 C609T polymorphism may play a pivotal role in tumor development. A significantly greater risk of hepatocellular carcinoma in association with the variant NQO1 C609T implies that this genotype may affect individual susceptibility to liver cancer.

14.5.2 Epoxide Hydrolase

Microsomal epoxide hydrolase (EPHX) is another important non-transferase detoxification enzyme. EPHX detoxifies low molecular weight chemicals such as butadiene, benzene, and styrene. In addition, EPHX also plays an important role in the activation and detoxification of polycyclic aromatic hydrocarbons present in tobacco smoking and grilled meat.

Interindividual variability attributes in part to the polymorphic expression of metabolic enzymes. EPHX gene at exons 3 and 4 exhibits two functional polymorphisms: EPHX T113C (Tyr → His variant in exon 3) and EPHX A139G (His → Arg variant in exon 4). These two coding region variants (T113C and A139G) affect the activity of EPHX enzyme, which may serve as molecular biomarkers to provide important information on carcinogenesis.

As a pathway to metabolize polycyclic aromatic hydrocarbons, microsomal EPHX T113C polymorphism may be considered as a genetic susceptibility factor for lung cancer and for the development of esophageal carcinogenesis. Individual susceptibility to colorectal cancer may be due in part to variations in detoxification capacity in the gastrointestinal tract. EPHX genetic polymorphisms may influence the levels of carcinogenic compounds present in colorectal. Consequently, EPHX genetic polymorphisms, T113C and A139G, appear to be an important factor for colorectal cancer risk.

EPHX has also been proposed to be a genetic factor for the development of chronic obstructive pulmonary disease. A significant correlation was found between EPHX genetic polymorphisms, T113C and A139G, and the individual's susceptibility to pulmonary disease and increased risk of developing aplastic anemia.

Bibliography

Abdel-Rahman SZ, Salama SA, Au WW, Hamada FA (2000) Role of polymorphic CYP2E1 and CYP2D6 genes in NNK-induced chromosome aberrations in cultured human lymphocytes. Pharmacogenetics 10(3):239–249

Angstadt AY, Berg A, Zhu J, Miller P, Hartman TJ, Lesko SM, Muscat JE, Lazarus P, Gallagher CJ (2013) The effect of copy number variation in the phase II detoxification genes UGT2B17 and UGT2B28 on colorectal cancer risk. Cancer 119(13):2477–2485

Autrup H (2000) Genetic polymorphisms in human xenobiotica metabolizing enzymes as susceptibility factors in toxic response. Mutat Res 464(1):65–76

Bajro MH, Josifovski T, Panovski M, Jankulovski N, Nestorovska AK, Matevska N, Petrusevska N, Dimovski AJ (2012) Promoter length polymorphism in UGT1A1 and the risk of sporadic colorectal cancer. Cancer Genet 205(4):163–167

Balliet RM, Chen G, Dellinger RW, Lazarus P (2010) UDP-glucuronosyltransferase 1A10: activity against the tobacco-specific nitrosamine, 4-(methylnitrosamino)-1-(3-pyridyl)-1-butanol, and a potential role for a novel UGT1A10 promoter deletion polymorphism in cancer susceptibility. Drug Metab Dispos 38(3):484–490

Boccia S, Persiani R, La Torre G, Rausei S, Arzani D, Gianfagna F, Romano-Spica V, D'Ugo D, Ricciardi G (2005) Sulfotransferase 1A1 polymorphism and gastric cancer risk: a pilot case-control study. Cancer Lett 229(2):235–243

Cascorbi I, Henning S, Brockmöller J, Gephart J, Meisel C, Müller JM, Loddenkemper R, Roots I (2000) Substantially reduced risk of cancer of the aerodigestive tract in subjects with variant—463A of the myeloperoxidase gene. Cancer Res 60(3):644–649

Chen CH (2012) Activation and detoxification enzymes: functions and implications. Springer Sciences, New York

Clapper ML (2000) Genetic polymorphism and cancer risk. Curr Oncol Rep 2(3):251–256

Daniels J, Kadlubar S (2013) Sulfotransferase genetic variation: from cancer risk to treatment response. Drug Metab Rev 45(4):415–422

Guillemette C, De Vivo I, Hankinson SE, Haiman CA, Spiegelman D, Housman DE, Hunter DJ (2006) Association of genetic polymorphisms in UGT1A1 with breast cancer and plasma hormone levels. Cancer Epidemiol Biomarkers Prev 10(6):711–714

Hein DW (2002) Molecular genetics and function of NAT1 and NAT2: role in aromatic amine metabolism and carcinogenesis. Mutat Res 506–507:65–77

Hisamuddin IM, Yang VW (2007) Genetic polymorphisms of human flavin-containing monooxygenase 3: implications for drug metabolism and clinical perspectives. Pharmacogenomics 8(6):635–643

Hong CC, Thompson HJ, Jiang C, Hammond GL, Tritchler D, Yaffe M, Boyd NF (2003) Val158Met polymorphism in catechol-O-methyltransferase gene associated with risk factors for breast cancer. Cancer Epidemiol Biomarkers Prev 12(9):838–847

Kadlubar FF, Badawi AF (1995) Genetic susceptibility and carcinogen-DNA adduct formation in human urinary bladder carcinogenesis. Toxicol Lett 82–83:627–632

Lee MS, Su L, Christiani DC (2010) Synergistic effects of NAT2 slow and GSTM1 null genotypes on carcinogen DNA damage in the lung. Cancer Epidemiol Biomarkers Prev 19(6):1492–1497

Li H, Fu WP, Hong ZH (2013) Microsomal epoxide hydrolase gene polymorphisms and risk of chronic obstructive pulmonary disease: a comprehensive meta-analysis. Oncol Lett 5 (3):1022–1030

Liu F, Luo L, Wei Y, Wang W, Li B, Yan L, Wen T (2013) A functional NQO1 609C>T polymorphism and risk of hepatocellular carcinoma in a Chinese population. Tumour Biol 34 (1):47–53

Mackenzie PI, Gregory PA, Lewinsky RH, Yasmin SN, Height T, McKinnon RA, Gardner-Stephen DA (2005) Polymorphic variations in the expression of the chemical detoxifying UDP glucuronosyltransferases. Toxicol Appl Pharmacol 207(2 Suppl):77–83

Matsui A, Ikeda T, Enomoto K, Nakashima H, Omae K, Watanabe M, Hibi T, Kitajima M (2000) Progression of human breast cancers to the metastatic state is linked to genotypes of catechol-O-methyltransferase. Cancer Lett 150(1):23–31

Merali Z, Ross S, Paré G (2014) The pharmacogenetics of carboxylesterases: CES1 and CES2 genetic variants and their clinical effect. Drug Metabol Drug Interact 29(3):143–151

Miller DP, De Vivo I, Neuberg D, Wain JC, Lynch TJ, Su L, Christiani DC (2003) Association between self-reported environmental tobacco smoke exposure and lung cancer: modification by GSTP1 polymorphism. Int J Cancer 104(6):758–763

Moslehi R, Chatterjee N, Church TR, Chen J, Yeager M, Weissfeld J, Hein DW, Hayes RB (2006) Cigarette smoking, N-acetyltransferase genes and the risk of advanced colorectal adenoma. Pharmacogenomics 7(6):819–829

Nebert DW, Petersen DD, Puga A (1991) Human AH locus polymorphism and cancer: inducibility of CYP1A1 and other genes by combustion products and dioxin. Pharmacogenetics 1(2):68–78

Phillips IR, Shephard EA (2019) Flavin-containing monooxygenase 3 (FMO3): genetic variants and their consequences for drug metabolism and disease. Xenobiotica 18:1–60

Sak K (2017) The Val158Met polymorphism in COMT gene and cancer risk: role of endogenous and exogenous catechols. Drug Metab Rev 49(1):56–83

Sak K, Everaus H (2016) Sulfotransferase 1A1 as a biomarker for susceptibility to carcinogenesis: from molecular genetics to the role of dietary flavonoids. Curr Drug Metab 17(6):528–541

Saldivar SJ, Wang Y, Zhao H, Shao L, Lin J, Spitz MR (2005) An association between a NQO1 genetic polymorphism and risk of lung cancer. Mutat Res 582(1-2):71–78

van der Hel OL, Peeters PH, Hein DW, Doll MA, Grobbee DE, Kromhout D, Bueno de Mesquita HB (2003) NAT2 slow acetylation and GSTM1 null genotypes may increase postmenopausal breast cancer risk in long-term smoking women. Pharmacogenetics 13(7):399–407

Zanger UM, Schwab M (2013) Cytochrome P450 enzymes in drug metabolism: regulation of gene expression, enzyme activities, and impact of genetic variation. Pharmacol Therapeut 138:103–141

van der Logt EM, Bergevoet SM, Roelofs HM, van Hooijdonk Z, te Morsche RH, Wobbes T, de Kok JB, Nagengast FM, Peters WH (2004) Genetic polymorphisms in UDP-glucuronosyltransferases and glutathione S-transferases and colorectal cancer risk. Carcinogenesis 25(12):2407–2415

van der Logt EM, te Morsche RH, Groenendaal N, Roelofs HM, de Metz M, van der Stappen JW, Nagengast FM, Peters WH (2009) Genetic polymorphism in UDP-glucuronosyltransferase 2B7 and colorectal cancer risk. Oncol Res 17(7):323–329

Vogel A, Ockenga J, Ehmer U, Barut A, Kramer FJ, Tukey RH, Manns MP, Strassburg CP (2002) Polymorphisms of the carcinogen detoxifying UDP-glucuronosyltransferase UGT1A7 in proximal digestive tract cancer. Z Gastroenterol 40(7):497–502

Wang YH, Lee YH, Tseng PT, Shen CH, Chiou HY (2008) Human NAD(P)H:quinone oxidoreductase 1 (NQO1) and sulfotransferase 1A1 (SULT1A1) polymorphisms and urothelial cancer risk in Taiwan. J Cancer Res Clin Oncol 134(2):203–209

Wu MT, Wang YT, Ho CK, Wu DC, Lee YC, Hsu HK, Kao EL, Lee JM (2003) SULT1A1 polymorphism and esophageal cancer in males. Int J Cancer 103(1):101–104

Yager JD (2000) Endogenous estrogens as carcinogens through metabolic activation. J Natl Cancer Inst Monogr 27:67–73

Zhang H, Liao LH, Liu SM, Lau KW, Lai AK, Zhang JH, Wang Q, Chen XQ, Wei W, Liu H, Cai JH, Lung ML, Tai SS, Wu M (2007) Microsomal glutathione S-transferase gene polymorphisms and colorectal cancer risk in a Han Chinese population. Int J Colorectal Dis 22(10):1185–1194

Zhang J, Schulz WA, Li Y, Wang R, Zotz R, Wen D, Siegel D, Ross D, Gabbert HE, Sarbia M (2003) Association of NAD(P)H: quinone oxidoreductase 1 (NQO1) C609T polymorphism with esophageal squamous cell carcinoma in a German Caucasian and a northern Chinese population. Carcinogenesis 24(5):905–909

Zhao W, Luo J, Cai X (2013) Association between microsomal epoxide hydrolase 1 polymorphisms and susceptibility to esophageal cancer: a meta-analysis. Tumour Biol 34(4):2383–2388

Chapter 15
Enzyme Polymorphisms Affecting Xenobiotic Toxicity

Foreign compounds to which humans are exposed undergo metabolic conversion mediated by metabolic enzymes, before they are excreted from the body. Activation enzymes catalyze functionalization of foreign compounds through oxidation, hydrolysis, and reduction reactions to form functionalized compounds. Functionalized compounds then proceed with further reactions catalyzed by detoxification enzymes to convert them into water-soluble and less active metabolites, thus facilitating their excretion from the organism. In many cases, during metabolic activation, foreign compounds are metabolized to form reactive intermediates that can bring about toxic effects by interacting with cellular components (proteins, lipids, and DNA).

As discussed in the previous chapter, advances in this area of research have revealed that many xenobiotic metabolic enzymes exhibit genetic polymorphisms. Genetic variability in the expression of metabolic enzymes is an important relevant feature associated with foreign compound-mediated biochemical and biomedical effects. Genetic polymorphisms underlying variations in enzymatic activity can also have a significant effect on an individual's susceptibility to diverse toxicity.

Individual differences resulting from enzyme polymorphisms in gene structure and tissue-specific expression can have a significant impact to the clearance of drugs and other chemicals and to the susceptibility of individuals to toxicity. Investigations of individual responsiveness to drugs or other chemicals have shown considerable deviations, in part due to variations in their metabolisms catalyzed by metabolic enzymes. Genetic polymorphisms play a crucial role in individual variations in response to foreign compound-mediated effects.

Moreover, the expression of foreign compound metabolic enzymes can also differ noticeably as a result of exposure to enzyme inducers. Both genetic polymorphisms and enzyme inducibility are of major significance. These two factors play important roles in the susceptibility of individuals to xenobiotic toxicities. The effects of

---

The original version of this chapter was revised. The correction to this chapter is available at https://doi.org/10.1007/978-3-030-41679-9_20

C.-H. Chen, *Xenobiotic Metabolic Enzymes: Bioactivation and Antioxidant Defense*,
https://doi.org/10.1007/978-3-030-41679-9_15

enzyme polymorphisms on xenobiotic metabolism and genetic polymorphisms on xenobiotic metabolic enzymes are discussed in the previous chapter. In the present chapter, the subject of genetic polymorphisms in the susceptibility of individuals to toxicity is discussed below.

15.1 Differential Susceptibility to Potentially Xenobiotic Toxicity

Genetic polymorphisms occur in many of the enzymes involved in metabolism of drugs and other xenobiotics. Genetic variations in the levels of expression and substrate specificity of metabolic enzymes can give rise to abnormal foreign compound metabolism. Such variations may account for the differential susceptibility to potentially toxic xenobiotics. Consequently, genetic differences in the expressions of genes coding for metabolic enzymes would be crucial factors in elucidating the susceptibility of drugs and other chemicals to the cytotoxicity or carcinogenesis.

15.1.1 Difference in Susceptibility to Drug Toxicity

A growing number of human genetic polymorphisms in drug-metabolizing enzymes have been characterized. Inheriting abnormal alleles of metabolic enzymes can alter drug metabolism. Such abnormal drug metabolism may be due to inherited factors which are (a) functionalization metabolism, such as inheriting abnormal alleles of cytochrome P450, or (b) conjugation metabolism, such as inheriting abnormal *N*-acetyltransferase.

There are interindividual and interethnic variabilities in the levels of activity of many drug-metabolizing enzymes. These variations could cause differences in susceptibilities to the effects and the toxicities of clinically used drugs and other chemicals. In many cases, polymorphisms in the gene encoding of these enzymes are the basis for this variability.

Individual adverse drug responses due to genetic risk factors can affect organs such as the liver or heart in a drug-dependent manner. Some of these are correlated with risk of cytotoxicity or carcinogenesis. Enzyme polymorphisms affecting the susceptibility to the toxicity of drugs such as tamoxifen, ipomeanol, aflatoxin, and tamoxifen are discussed below.

15.1.2 Susceptibility to Diverse Toxicities

Polymorphic variants in enzyme genes may lead to variations in the extent of activation, thereby influencing individual sensitivity toward xenobiotic toxicity. For instances, majority of nicotine is oxidized by CYP450 to nicotine-iminium ions, which are further metabolized to cotinine by aldehyde oxidase and then to hydrolated products. Moreover, genetic polymorphisms of CYP450s may lead to interindividual differences in CYP450-mediated oxidation of nitrosamines. CYP450s are also involved in metabolizing styrene to styrene 7,8-oxide, which is mainly hydrolyzed to styrene glycol by epoxide hydrolase.

Differences in the levels of expression and catalytic activities of activation enzymes such as CYP450s and detoxification enzymes such as GSTs may influence individual sensitivity toward tobacco smoke toxicity. Metabolism of benzo[a]pyrene in tobacco smoke involves activation metabolism catalyzed by CYP450, epoxide hydrolase, and prostaglandin-H synthase, to form dihydrodiol epoxide, which is capable of covalently binding to DNA. The detoxification of benzo[a]pyrene is then catalyzed by glutathione S-transferases (GST).

15.2 Enzyme Polymorphisms Affecting Xenobiotic Metabolisms

Genetic variability in the expression of metabolic enzymes is an important relevant feature associated with foreign compound-mediated biochemical and biomedical effects. Genetic polymorphisms underlying variations in enzymatic activity can have a significant effect on an individual's susceptibility to diverse toxicity. The differences in individual sensitivity to xenobiotic metabolites could explain many observed differential reactions to foreign compounds among individuals. There has been a great deal of interest in the role of genetic polymorphisms of metabolic enzymes in the pathogenesis of adverse responses.

Activation enzymes catalyze functionalization reactions, resulting in forming reactive intermediates. Many toxicants or carcinogens are metabolized to active intermediates prior to eliciting their toxic effects. For instance, polymorphic variations in CYP450 genes may lead to variations in the susceptibility to foreign compound toxicity. Other activation enzymes, alcohol dehydrogenase (ADH) and aldehyde dehydrogenase (ALDH), are responsible for the oxidative metabolism of ethanol. Like CYP450s, ADH and ALDH are also polymorphic in humans. ADH oxidizes ethanol to acetaldehyde, and ALDH oxidizes acetaldehyde to acetate, which is further metabolized to CO_2 and H_2O.

Detoxification enzymes, such as glutathione S-transferases (GST) and UDP-glucuronosyl transferase (UGT), conjugate functionalized foreign compounds to water-soluble derivatives, thus detoxifying and facilitating their excretion from the body. Genetic polymorphisms in GST and UGT may result in impaired detoxification, which may have significant impacts on an individual's susceptibility to foreign compound-mediated toxicities, such as GST on tobacco smoke and UGT on tamoxifen.

The susceptibility to foreign compound toxicity mediated by genetic polymorphisms of metabolic enzymes, including CYP450, ADH and ALDH of activation enzymes, and GST and UGT of detoxification enzymes, is discussed below.

15.2.1 Cytochrome P450

There are significant differences in the levels of expression of CYP450 isozymes between individuals, as evaluated by the analysis of human liver samples in clinical pharmacological studies. Among them, a noticeable human CYP450-dependent chemical metabolism is carried out by polymorphic enzymes. For instance, genetic polymorphisms of CYP450 isozymes, 2A6, 1A1, and 2E1, have been investigated to examine their impacts on individual susceptibility to the toxicity of foreign compounds. Such studies are briefly discussed below.

15.2.1.1 CYP2A6 Polymorphisms Affecting Nicotine Metabolism

A high expression of CYP450 activity may result in an undesired increase in the rate of activation of foreign compounds, leading to overproduction of reactive intermediates and metabolites. Metabolisms mediated by high activity of CYP450 have been implicated in the produced toxicity and carcinogenicity of certain foreign compounds. CYP2A6 is involved in the metabolic activation of nicotine.

Enzyme polymorphisms have been reported to influence the rate of nicotine elimination from the body. Figure 15.1 illustrates polymorphic variants in CYP2A6 that affect nicotine metabolism. The produced nicotine-iminium ion can react with DNA, leading to higher lung cancer risk.

Interindividual and interethnic variability in the level of CYP2A6 activity is attributed to polymorphisms in the CYP2A6 gene. CYP2A6 is present predominantly in the liver. Over 22 different alleles have been characterized. This enzyme also activates a number of structurally unrelated chemical carcinogens including aflatoxin B1, halothane, and valproic acid.

CYP2A6 gene is highly polymorphic: it contains more than a dozen different alleles including CYP2A6∗1A and CYP2A6∗1B. A relatively high allele frequency of the CYP2A6 gene deletion has been reported in Asian population. This population has a generally reduced CYP2A6 activity. CYP2A6∗1A genotype that causes a decreased expression of CYP2A6 has been shown to have a significant association with decreased risk for lung cancer. This observed decreased risk is due to a lower rate of formation of carcinogenic products by a less active CYP2A6∗1A.

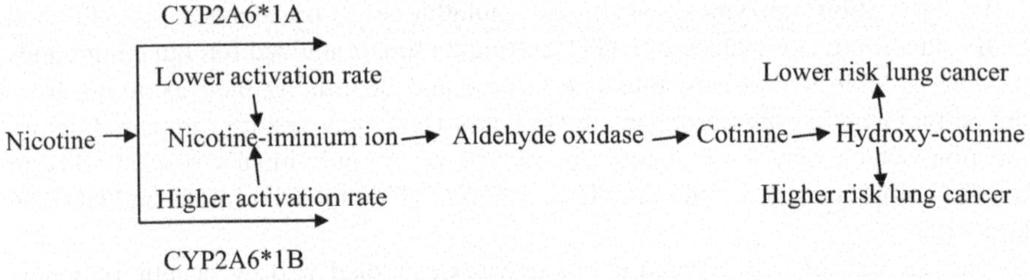

Fig. 15.1 CYP2A6 polymorphisms affecting nicotine metabolism

In contrast, individuals with the CYP2A6*1B allele have higher CYP2A6 expression as well as catalytic activity. These individuals are likely to have increased risk for lung cancer. These reports demonstrate that genetic polymorphisms of CYP2A6 genotype can influence the susceptibility of individuals to lung cancer.

15.2.1.2 CYP1A1 Polymorphisms Affecting Benzo[a]pyrene Metabolism

Enzyme polymorphism also plays an important role in the differences between individuals in response to the toxicity of polycyclic aromatic hydrocarbons. Polymorphic variants in CYP450 gene, particularly CYP1A1, have an effect on the metabolism of these chemicals, thereby influencing the susceptibility of individuals toward their toxicity.

The frequency of the genotypes, Ile/Ile, Ile/Val, and Val/Val, has been assessed to evaluate the relationship between CYP1A1 polymorphism and individual sensitivity to the toxicity of polycyclic aromatic compounds, which may affect the incidence of lung cancer. A higher level of DNA adducts has been detected in lymphocytes of tobacco smokers with specific polymorphism of CYP1A1 gene.

Figure 15.2 illustrates polymorphic variants in CYP1A1 affecting activation of benzo[a]pyrene. Specific genotype Val/Val for CYP1A1 was found to be related to the incidence of esophageal carcinoma, especially in heavy smokers. CYP1A1 Ile/Val polymorphism was also reported to associate with an elevated risk of lung cancer among Japanese population.

15.2.1.3 CYP2E1 Polymorphisms Affecting Nitrosamine and Styrene Metabolism

CYP2E1, another polymorphic variant in CYP450 genes, is a key enzyme in the metabolic activation of a variety of toxicants, including benzene, vinyl chloride, and trichloroethylene. CYP2E1 is also a key enzyme involved in the metabolism of

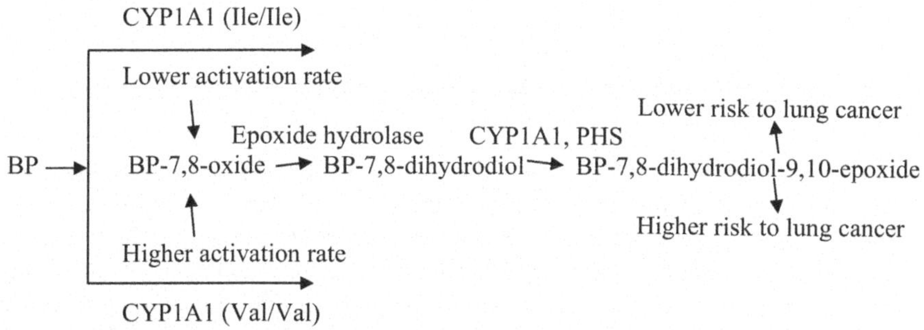

Fig. 15.2 CYP1A1 polymorphisms affecting benzo[a]pyrene (BP) metabolism

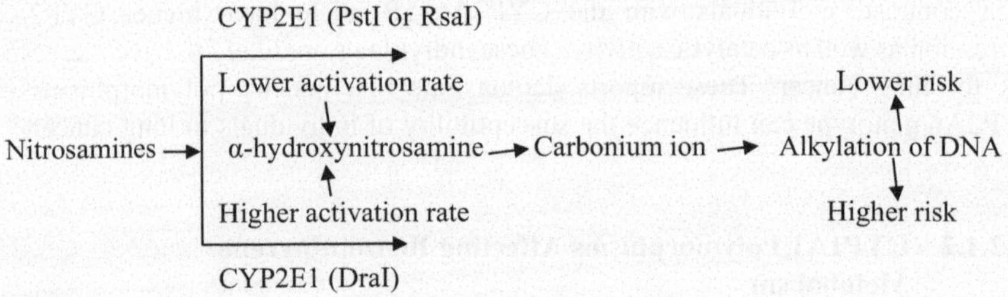

Fig. 15.3 CYP2E1 polymorphisms affecting nitrosamine activation

nitrosamines such as 4-(methylnitrosamino)-1-(3-pyridyl)-1-butanone and *N*-′-nitrosonornicotine in tobacco smoke. Figure 15.3 illustrates polymorphic variants in CYP2E1 affecting metabolism of nitrosamines. The generated reactive intermediate, carbonium ion, can react with DNA.

The variant allele of three polymorphisms (PstI, RsaI, and DraI) has been assessed. Some studies have demonstrated an association of the CYP2E1 gene with the susceptibility of individuals to lung cancer. Genetic polymorphism of CYP2E1 DraI was found to associate with an increased risk of lung cancer in the Japanese.

Styrene is a monocyclic aromatic hydrocarbon compound, which is used in the production of rubber, resin, and fiberglass. It is absorbed via the lung and through skin contact. Styrene is primarily metabolized in the liver. Styrene toxicity is generally induced by the production of active metabolic intermediates such as styrene 7,8-oxide.

The primary target for styrene toxicity appears to be the central and peripheral nervous systems. Individual differences in urinary metabolite concentrations have been associated with genetic polymorphism of styrene-metabolizing enzymes such as CYP2E1. Urinary phenyl hydroxyethyl mercapturic acids as minor metabolite of styrene are associated with GST polymorphisms. CYP2E1 is also one of the enzymes that metabolize ethanol to acetaldehyde. Genetic polymorphisms of CYP2E1 have been identified and linked to altered susceptibility of individuals to alcoholism.

15.2.2 Alcohol Dehydrogenase and Aldehyde Dehydrogenase

Alcoholism is a typical example involving genetic polymorphisms that affect the expression of metabolizing enzymes. The kinetics of ethanol absorption and elimination is influenced by genetic factors. The liver enzymes, alcohol dehydrogenase (ADH) and aldehyde dehydrogenase (ALDH), are two important enzymes involving alcohol metabolism. ADH and ALDH are responsible for the oxidative metabolism of ethanol. ADH catalyzes oxidation of ethanol to acetaldehyde, and ALDH oxidizes acetaldehyde to acetate, which is further metabolized to CO_2 and H_2O.

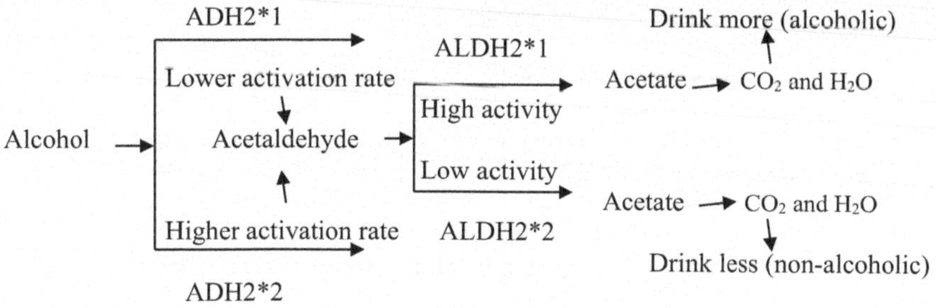

Fig. 15.4 ADH and ALDH polymorphisms affecting alcohol metabolisms

ADH and ALDH enzymes are polymorphic in humans. Genetic polymorphisms of ADH and ALDH have an influence on alcoholism susceptibility. ADH is present as a polygene family, which has seven genes including ADH2∗1, ADH2∗2, and ADH2∗3 (family, 1, 2, and 3; allele, 1, 2, and 3, respectively). They are functionally polymorphic.

Genetic variations in ADH and ALDH among individuals in response to alcohol have been reported to contribute to differences in susceptibility to alcohol abuse and dependence, particularly, the functional genetic variants of ADH that exhibit low alcohol oxidizing activity. The genetic variants of ALDH that exhibit low acetaldehyde oxidizing activity protect against heavy drinking and alcoholism.

Figure 15.4 shows polymorphic variants in ADH and ALDH affecting alcohol metabolism. The alcoholics were found to have significantly higher frequencies of ALDH2∗1 allele than did the healthy subjects.

15.2.3 Glutathione-S-Transferase

Genetic polymorphisms of human glutathione S-transferase (GST) consist of multiplicities of GST such as GSTM1 and GSTT1 genes. Genetic polymorphisms in GST may result in impaired detoxification and therefore have significant impacts on an individual's susceptibility to foreign compound-mediated toxicities. For example, blood cultures obtained from GSTT1- and GSTM1-null individuals were found to exhibit an increase in the sensitivity to various toxins.

15.2.3.1 GST Polymorphisms Affecting Susceptibility to DNA Damages

Individuals who carry GSTM1-null genotype were reported to have a higher susceptibility to DNA damage induced by tobacco smoke than those with GSTM1-positive genotype. A higher level of DNA adducts has been detected in lymphocytes of tobacco smokers who lack the GSTM1 gene.

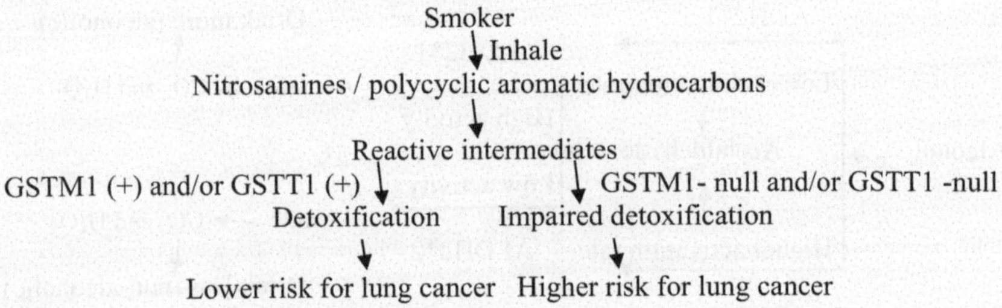

Fig. 15.5 GST polymorphisms affecting smoker risk to cancer

Several studies have also revealed an association of GSTM1- and GSTT1-null genotypes with an increasing risk of lung cancer. Figure 15.5 shows polymorphic variants in GST affecting smoker risk to cancer. Individuals lacking both GSTM1 and GSTT1 activity were also found to be at increased risk for gastric cancer.

15.2.3.2 GTS Polymorphisms Affecting Susceptibility to Leukemia

Glutathione S-transferase enzyme polymorphisms were found to associate with the susceptibility to acute leukemia in adults. GSTT1-null variants were reported to increase leukemia risk. Oxidative stress in hematopoietic stem cells has been implicated as an important etiological factor in acute myeloid leukemia. Industrial solvent benzene is a well-recognized leukemogen (substance that produces leukemia). Genetic variants in its metabolic pathway can modulate the risk of leukemia following its exposure.

15.2.4 UDP-Glucuronosyl Transferase

UDP-glucuronosyltransferases convert foods, drugs, and environmental chemicals to inactive water-soluble glucuronides. The expression of UGT1 and UGT2 gene families is regulated in a tissue-specific fashion. Tissue-specific and interindividual polymorphic regulation of UGT1A and UGT2B genes in small intestine was reported to implicate as molecular biological determinant contributing to interindividual xenobiotic metabolism in human.

15.2.4.1 UGT Polymorphisms Affecting Tamoxifen Metabolism

As a selective estrogen receptor modulator which blocks the effects of estrogen in the breast tissue, tamoxifen is an important antiestrogenic drug used for the treatment of breast cancer. Tamoxifen undergoes oxidation reaction catalyzed by CYP2D6 to form reactive metabolites, 4-hydroxy-tamoxifen (4HT) and 4-hydroxy-N-desmethyl tamoxifen (endoxifen), which negate their antiestrogenic properties. These reactive intermediates are then detoxified through glucuronidation by UDP-glucuronosyl transferases (UGT). Three UGTs that are encoded by polymorphic genes, UGT2B15*2, UGT2B7*2, and UGT1A8*3, could have the potential to modify the effectiveness of detoxifying tamoxifen.

Nevertheless, investigations of functional polymorphisms in key tamoxifen-metabolizing enzymes revealed that UGT polymorphisms were not associated with breast cancer recurrence risk. No association was found between genetically modified UGT activity and the rate of breast cancer recurrence. These findings suggested that the genotyping of key metabolic enzyme polymorphisms could not predict response to tamoxifen therapy.

15.2.4.2 UGT Polymorphisms Affecting Acetaminophen Metabolism

Acetaminophen is used to temporarily relieve ache and fever due to headache. Detoxifying metabolism of acetaminophen is conjugated with glucuronic acid and sulfate before its excretion from the body. The glucuronidation is thought to occur mainly catalyzed by UGT. Interindividual variation in acetaminophen glucuronidation is attributed in part to polymorphisms in enzyme UGT1A (UGT family 1 and subfamily A). Acetaminophen glucuronidation was found to be significantly influenced by UGT2B15 polymorphisms.

15.3 Species Differences Affecting Foreign Compound Toxicity

Variations in detoxification enzyme activity are often the basis for species differences in an organism's susceptibility to foreign compound-mediated toxicity. A number of studies have been reported to support the proposal that metabolic differences are the underlying cause of species variation in cytotoxicity and carcinogenicity.

The susceptibility to aflatoxin toxicity is different between humans and mice and rats. Resistance to tamoxifen toxicity was reported in humans, but not in rats. Humans and rodents were also found to exhibit a different toxic response to 4-ipomeanol, a pulmonary pre-toxin isolated from sweet potatoes infected with fungus. Such differences are mainly attributed to species variations in the activities of metabolic enzymes that catalyze the metabolisms of drugs and other chemicals. Species differences in the susceptibility to aflatoxin-, tamoxifen-, and 4-ipomeanol-induced toxicities are discussed below.

15.3.1 Susceptibility to Aflatoxin Toxicity in Humans, But Not in Mice

Aflatoxin B_1, a member of a group of mycotoxins, is a major risk factor for hepatocellular carcinoma in humans. In humans, enzyme CYP3A4 is abundant in the hepatic and the intestines. Metabolic activation of aflatoxin B_1 catalyzed by CYP3A4 forms a carcinogenic reactive intermediate, aflatoxin-8,9-epoxide, which can react with DNA to form a DNA adduct.

During detoxification metabolism, aflatoxin-8,9-epoxide undergoes conjugation reaction catalyzed by GSTA3 enzyme to form glutathione-conjugated aflatoxin derivative, thus facilitating its excretion from the body. Thus, GSTA3 plays a crucial role in protection against aflatoxin B_1 toxicity. Nevertheless, due to relatively low expression of GSTA3 activity for aflatoxin-8,9-epoxide, humans are highly susceptible to toxic and carcinogenic effects of aflatoxin B_1.

Conversely, mice are resistant to aflatoxin B_1 carcinogenesis because of a high expression of GSTA3 in the mouse liver. Glutathione conjugation of aflatoxin-8,9-epoxide confers intrinsic resistance to aflatoxin carcinogenesis, causing no occurrence of liver cancer in mice.

The resistance of mice to aflatoxin B_1 carcinogenesis due to a high expression of mGSTA3 was further confirmed by investigating knockout mice.

Such species difference in the susceptibility to aflatoxin toxicity between humans and mice is presented in Fig. 15.6. The deletion of mGSTA3 gene in knockout mice was found to have more than 100-fold of aflatoxin-DNA adduct in the liver than wild-type mice, and knockout mice responded to aflatoxin B_1 toxicity in a manner similar to humans.

15.3.2 Resistance to Tamoxifen Toxicity in Humans, But Not in Rats

Metabolic activation of tamoxifen catalyzed by CYP3A4 involves hydroxylation, which results in producing metabolic intermediates such as α-hydroxytamoxifen

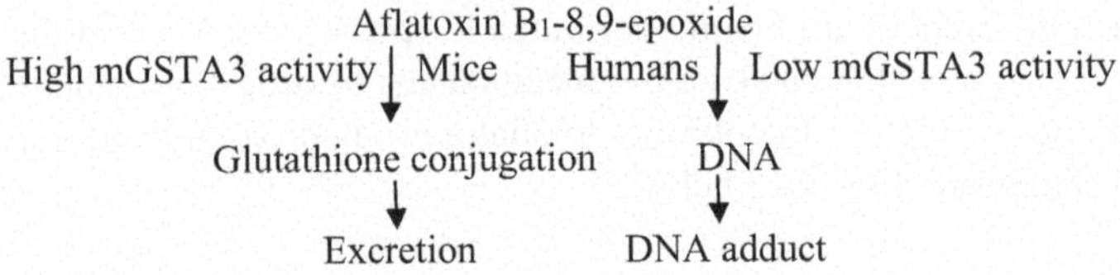

Fig. 15.6 Susceptibility to aflatoxin B_1 toxicity in humans, but not in mice

derivatives. During detoxification metabolism, the glucuronylation pathway competes with the sulfonation pathway. In the glucuronylation pathway, a conjugation reaction catalyzed by UGT forms inactive tamoxifen glucuronidate, while in the sulfonation pathway, a conjugation reaction catalyzed by sulfotransferase (ST) produces tamoxifen o-sulfonate.

Humans exhibit a much higher activity of UGT than rats and have more effective glucuronylation of α-hydroxytamoxifen in liver microsomes than rats. The high UGT activity plays a key role for humans to defend against the toxicity of tamoxifen. Such a high detoxification rate in liver microsomes has been reported to correlate with a low risk of tamoxifen in causing liver carcinoma in humans.

In contrast, tamoxifen has been shown to be a potent hepatocarcinogen in rats. Rats display a low activity of UGT, but a high activity of ST. As a result, α-hydroxytamoxifen in rats prefers a conjugation reaction through sulfonation to glucuronidation. Although sulfonation conjugation usually is a detoxification reaction, tamoxifen o-sulfonate subsequently releases its sulfonate group to produce reactive carbocation, which can react with DNA to form DNA adducts. Consequently, DNA adducts were detected in hepatocytes of rats treated with tamoxifen. Tamoxifen is hence a liver carcinogen in rats.

Figure 15.7 presents species differences in tamoxifen detoxification metabolism between humans and rats. The figure reveals that a higher activity for glucuronylation of α-hydroxytamoxifen in combination with a lower rate of sulfonation protects humans from forming tamoxifen-DNA adducts. As a result, humans exhibit a low risk to tamoxifen-associated liver carcinoma, and DNA adducts were not detected in tamoxifen-treated human hepatocytes.

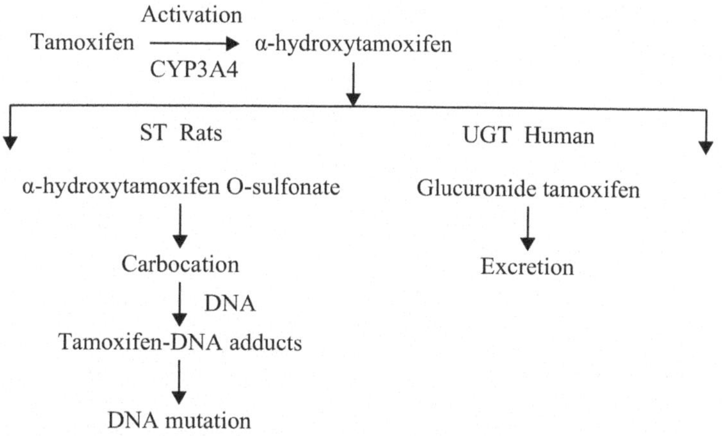

Fig. 15.7 Susceptibility to tamoxifen toxicity in rats, but not in humans

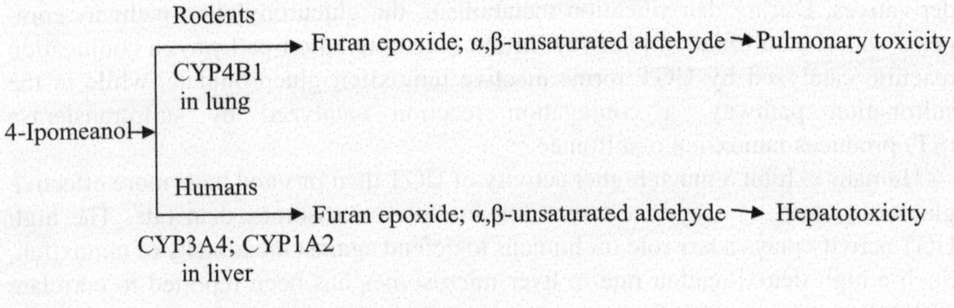

Fig. 15.8 4-Ipomeanol toxicity response: humans versus rodents

15.3.3 Humans Versus Rodents in Response to 4-Ipomeanol Toxicity

4-Ipomeanol, a pulmonary toxin, was reported to induce pulmonary toxicity in the lungs of rodents and cattle. Metabolic activation of 4-ipomeanol catalyzed by CYP4B1 produces a highly reactive furan epoxide, which induces toxicity in the lung of these animals. In contrast, in humans, 4-ipomeanol is metabolically activated in the liver by hepatic enzymes, CYP3A4 and CYP1A2. The produced reactive furan epoxide causes selective hepatotoxicity in humans, while the risk for developing pulmonary toxicity in humans is minimal.

Figure 15.8 illustrates distinctive responses to 4-ipomeanol toxicity between humans and rodents. The figure demonstrates the deviation in response to 4-ipomeanol toxicity between humans and rodents: not only the different sites of metabolism but also the distinctive CYP450 enzymes present in metabolic activation of 4-ipomeanol.

Bibliography

Ahern TP, Christensen M, Cronin-Fenton DP, Lunetta KL, Søiland H, Gjerde J, Garne JP, Rosenberg CL, Silliman RA, Sørensen HT, Lash TL, Hamilton-Dutoit S (2011) Functional polymorphisms in UDP-glucuronosyl transferases and recurrence in tamoxifen-treated breast cancer survivors. Cancer Epidemiol Biomarkers Prev 20(9):1937–1943

Alvarez-Diez TM, Zheng J (2004) Mechanism-based inactivation of cytochrome P450 3A4 by 4-ipomeanol. Chem Res Toxicol 17:150–157

Archer MC (1981) Reactive intermediates from nitrosamines. Adv Exp Med Biol 136:1027–1035

Ariyoshi N, Miyamoto M, Umetsu Y, Kunitoh H, Dosaka-Akita H, Sawamura Y, Yokota J, Nemoto N, Sato K, Kamataki T (2002) Genetic polymorphism of CYP2A6 gene and tobacco-induced lung cancer risk in male smokers. Cancer Epidemiol Biomarkers Prev 11:890–894

Baer BR, Rettie AE, Henne KR (2005) Bioactivation of 4-ipomeanol by CYP4B1: adduct characterization and evidence for an enedial intermediate. Chem Res Toxicol 18:855–864

Baum M, Amin S, Guengerich FP, Hecht SS et al (2001) Metabolic activation of benzo[c] phenanthrene by cytochrome P450 enzymes in human liver and lung. Chem Res Toxicol 14:686–693

Boocock DJ, Maggs JL, Brown K et al (2000) Major inter-species differences in the rates of O-sulphonation and O-glucuronylation of alpha-hydroxytamoxifen in vitro: a metabolic disparity protecting human liver from the formation of tamoxifen-DNA adducts. Carcinogenesis 21:1851–1858

Bourian M, Gullstén H, Legrum W (2000) Genetic polymorphism of CYP2A6 in the German population. Toxicology 144(1–3):129–137

Brockmöller J, Cascorbi I, Henning S, Meisel C, Roots I (2000) Molecular genetics of cancer susceptibility. Pharmacology 61(3):212–227

Chen CH (2012) Activation and detoxification enzymes: functions and implications. Springer Sciences, New York

Czerwinski M, McLemore TL, Philpot RM et al (1991) Metabolic activation of 4-ipomeanol by complementary DNA-expressed human cytochromes P-450: evidence for species-specific metabolism. Cancer Res 51:4636–4638

Daly AK (2013) Pharmacogenomics of adverse drug reactions. Genome Med 5(1):5

Falzon M, McMahon JB, Schuller HM et al (1986) Metabolic activation and cytotoxicity of 4-Ipomeanol in human non-small cell. lung cancer lines. Cancer Res 46:3484–3489

He X, Feng S (2015) Role of metabolic enzymes P450 (CYP) on activating procarcinogen and their polymorphisms on the risk of cancers. Curr Drug Metab 16(10):850–863

Ilic Z, Crawford D, Vakharia D et al (2010) Glutathione-S-transferase A3 knockout mice are sensitive to acute cytotoxic and genotoxic effects of aflatoxin B1. Toxicol Appl Pharmacol 242:241–246

Johansson I, Ingelman-Sundberg M (2011) Genetic polymorphism and toxicology—with emphasis on cytochrome P450. Toxicol Sci 120(1):1–13

Jowsey IR, Jiang Q, Itoh K et al (2003) Expression of the aflatoxin B1-8,9-epoxide-metabolizing murine glutathione S-transferase A3 subunit is regulated by the Nrf2 transcription factor through an antioxidant response element. Mol Pharmacol 64:1018–1028

Kato S, Shields PG, Caporaso NE et al (1994) Analysis of cytochrome P450 2E1 genetic polymorphisms in relation to human lung cancer. Cancer Epidemiol Biomarkers Prev 3:515–518

Kilburn L, Okcu MF, Wang T, Cao Y, Renfro-Spelman A, Aldape KD, Gilbert MR, Bondy M (2010) Glutathione S-transferase polymorphisms are associated with survival in anaplastic glioma patients. Cancer 116(9):2242–2249

Kim KW (2015) Effects of styrene-metabolizing enzyme polymorphisms and lifestyle behaviors on blood styrene and urinary metabolite levels in workers chronically exposed to styrene. Toxicol Res 31(4):355–361

Kim SY, Laxmi YR, Suzuki N et al (2005) Formation of tamoxifen-DNA adducts via O-sulfonation, not O-acetylation, of alpha-hydroxytamoxifen in rat and human livers. Drug Metab Dispos 33:1673–1678

Kiss I, Sándor J, Pajkos G, Bogner B et al (2000) Colorectal cancer risk in relation to genetic polymorphism of cytochrome P4501A1, 2E1 and glutathione-S-transferase M1 enzymes. Anticancer Res 20:519–522

Kiyohara C (2000) Genetic polymorphism of enzymes involved in xenobiotic metabolism and the risk of colorectal cancer. J Epidemiol 10(5):349–360

Kondraganti SR, Fernandez-Salguero P et al (2003) Polycyclic aromatic hydrocarbon-inducible DNA adducts: evidence by 32P-postlabeling and use of knockout mice for Ah receptor-independent mechanisms of metabolic activation in vivo. Int J Cancer 103:5–11

Li TK (2000) Pharmacogenetics of responses to alcohol and genes that influence alcohol drinking. J Stud Alcohol 61:5–12

Maezawa Y, Yamauchi M, Toda G, Suzuki H, Sakurai S (1995) Alcohol-metabolizing enzyme polymorphisms and alcoholism in Japan. Alcohol Clin Exp Res 19(4):951–954

Morgan GJ, Smith MT (2002) Metabolic enzyme polymorphisms and susceptibility to acute leukemia in adults. Am J Pharmacogenomics 2(2):79–92

Neafsey P, Ginsberg G, Hattis D et al (2009) Genetic polymorphism in CYP2E1: population distribution of CYP2E1 activity. J Toxicol Environ Health B Crit Rev 12:362–388

Nebert DW, McKinnon RA, Puga A (1996) Human drug-metabolizing enzyme polymorphisms: effects on risk of toxicity and cancer. DNA Cell Biol 15(4):273–280

Nimura Y, Yokoyama S, Fujimori M et al (1997) Genotyping of the CYP1A1 and GSTM1 genes in esophageal carcinoma patients with special reference to smoking. Cancer 80:852–857

Oscarson M (2001) Genetic polymorphisms in the cytochrome P450 2A6 (CYP2A6) gene: implications for interindividual differences in nicotine metabolism. Drug Metab Dispos 29:91–95

Oscarson M, McLellan RA, Gullstén H et al (1999) Identification and characterisation of novel polymorphisms in the CYP2A locus: implications for nicotine metabolism. FEBS Lett 460:321–327

Palma S, Cornetts T, Padua L et al (2007) Influence of glutathione S-transferase polymorphisms on genotoxic effects induced by tobacco smoke. Mutat Res 633:1–12

Phillips DH, Carmichael PL, Hewer A et al (1996) Activation of tamoxifen and its metabolite alpha-hydroxytamoxifen to DNA-binding products: comparisons between human, rat and mouse hepatocytes. Carcinogenesis 17:89–94

Pirmohamed M, Park BK (2003) Cytochrome P450 enzyme polymorphisms and adverse drug reactions. Toxicology 192(1):23–32

Rothman N, Shields PG, Poirier MC et al (1995) The impact of glutathione s-transferase M1 and cytochrome P450 1A1 genotypes on white-blood-cell polycyclic aromatic hydrocarbon-DNA adduct levels in humans. Mol Carcinog 14:63–68

Sandi L, Navarro SL, Chen Y, Li L, Li SS, Chang JL, Schwarz Y, King IB, Potter JD, Bigler J, Lampe JW (2011) UGT1A6 and UGT2B15 polymorphisms and acetaminophen conjugation in response to a randomized, controlled diet of select fruits and vegetables. Drug Metab Dispos 39 (9):1650–1657

Sgambato A, Campisi B, Zupa A et al (2002) Glutathione S-transferase (GST) polymorphisms as risk factors for cancer in a highly homogeneous population from southern Italy. Anticancer Res 22:3647–3652

Shupe T, Sell S (2004) Low hepatic glutathione S-transferase and increased hepatic DNA adduction contribute to increased tumorigenicity of aflatoxin B1 in newborn and partially hepatectomized mice. Toxicol Lett 148:1–9

Silveira VS, Canalle R, Scrideli CA, Queiroz RG, Lopes LF, Tone LG (2012) CYP3A5 and NAT2 gene polymorphisms: role in childhood acute lymphoblastic leukemia risk and treatment outcome. Mol Cell Biochem 364(1–2):217–223

Strassburg CP, Kneip S, Topp J, Obermayer-Straub P, Barut A, Tukey RH, Manns MP (2000) Polymorphic gene regulation and interindividual variation of UDP-glucuronosyltransferase activity in human small intestine. J Biol Chem 275(46):36164–36171

Wang J, Pitarque M, Ingelman-Sundberg M (2005) 3′-UTR polymorphism in the human CYP2A6 gene affects mRNA stability and enzyme expression. Biochem Biophys Res Commun 340:491–497

Watanabe J, Hayashi S, Kawajiri K (1994) Different regulation and expression of the human CYP2E1 gene due to the RsaI polymorphism in the 5′-flanking region. J Biochem 116:321–326

Chapter 16
Inducibility of Foreign Compound Metabolic Enzymes

Inducibility of metabolic activation and detoxification enzymes is another important feature of xenobiotic metabolism, besides genetic polymorphisms. Metabolic enzyme induction refers to an increase in the rate of xenobiotic metabolism mediated by increasing transcription of mRNA which encodes the genes for metabolic enzymes. Generally, enzyme induction is a process in which a compound enhances and induces the expression of enzyme. Extensive research has revealed that some small molecules or compounds are able to modulate xenobiotic metabolic enzymes.

Metabolic enzyme induction occurs when a molecule or compound causes an increase or decrease in the activity of enzymes, thereby affecting the metabolism of foreign compounds, which are catalyzed by metabolic enzymes. An increased rate of detoxification for the parent compound would result in a lower potency and reduced effect of foreign compounds. Meanwhile, the relative rates of activation and detoxification activities are even more important in the determination of the toxic effects of reactive intermediates and metabolites that are generated from xenobiotic metabolism.

Enzyme inducibility is of major significance because it plays a role in the susceptibility of individuals to foreign compound-mediated toxicities. Strategies for protecting cells from initiation toxic events include (a) a decrease in the expression of activation enzymes responsible for generating reactive species and (b) an increase in the activities of detoxification enzymes that detoxify electrophilic reactive intermediates, including reactive oxygen species and free radicals known to intervene in regular cellular processes.

It is also essential to realize that the consequence of enzyme induction or inhibition is dependent on not only the specific effect on certain activation enzymes or detoxification enzymes but also the balance between the rates of activation and detoxification. Induction of activation enzymes leads to a higher expression of enzyme activity, resulting in an enhancement in the activation rate and the production of potential reactive intermediates. If the generated reactive intermediates are not immediately removed from the body, the induction of activation enzymes would enhance the toxic effects of foreign compounds.

C.-H. Chen, *Xenobiotic Metabolic Enzymes: Bioactivation and Antioxidant Defense*,
https://doi.org/10.1007/978-3-030-41679-9_16

In contrast, the induction of detoxification enzymes leads to a higher expression of enzyme activity and an increase in the rate of detoxifying reactions. Thus, to a foreign compound whose toxicity is derived from reactive intermediates mediated by metabolic activation, inhibition of activation enzyme or induction of detoxification enzymes would decrease xenobiotic toxic effects. It should be pointed out that the effects of metabolic activation and detoxification on xenobiotic toxicity depend on not only the relative activities of these enzymes but also the nature of the foreign compounds.

Enzyme modulators are small molecules or compounds that cause induction or inhibition effects on the activities of activation and detoxification enzymes. The action of such a small molecule or compound involves covalently or non-covalently binding to an enzyme, thereby changing enzyme activity in a manner that either increases or decreases its catalytic capacity. To exert such action, it is essential that modulators are able to penetrate across cell membranes. Moreover, enzymatic catalysis processes are also mediated by functional groups present in the amino acid side chains.

The structure of enzyme is required to stabilize the conformation of the active site for achieving enzymatic function. The active site of an enzyme is usually larger than the substrate, and the substrate is partially surrounded by the active site in most cases. An efficient condition for an enzyme catalytic action is to have the activity of enzyme proportional available whenever it is needed, so if the cells accumulate toxic metabolites in large quantity, they are able to dispose them by activating appropriate metabolic enzymes.

16.1 Inducibility of Metabolic Enzymes

In order to exert toxic and neoplastic effects, most toxic or carcinogenic foreign compounds require activation to reactive intermediates catalyzed by activation enzymes such as cytochromes P450. To counter the toxicity and carcinogenesis of foreign compounds, two major lines of defense are detoxification enzymes that serve as indirect antioxidants and intracellular antioxidants such as glutathione that serve as direct antioxidants. These two defenses play a prominent role in providing cells protection against cytotoxicity and carcinogenesis derived from xenobiotic metabolism.

In the line of defense with detoxification enzymes, reactive intermediates generated from activation metabolism are susceptible to metabolic conjugation catalyzed by detoxification enzymes, such as glutathione S-transferases, UDP-glucuronosyltransferases, and NAD(P)H:quinone reductase. Extensive research has demonstrated the inducibility of these detoxification enzymes by small molecules or compounds such as isothiocyanates and 1,2-dithio-3-thiones. The induction of detoxification genes encoding for proteins offers protection against the damages of oxidative stress and inflammation.

Accordingly, induction of detoxification enzymes has been considered as an effective strategy for combating the risk of chronic degenerative diseases and various disease conditions. The family of inducible detoxifying enzymes is diverse. Besides glutathione *S*-transferases and UDP-glucuronosyltransferase, inducible enzymes also includes NAD(P)H:quinone reductase, epoxide hydrolase, catalase, superoxide dismutase, and heme oxygenase 1. Heme oxygenase 1 catalyzes the conversion of heme to biliverdin, which is reduced enzymatically to bilirubin, a potent antioxidant. Induction of heme oxygenase may also protect cells against oxidative damages.

16.2 A Variety of Metabolic Enzyme Inducers

Detoxification enzymes such as glutathione transferase (GTS), NAD(P)H:quinone reductase (QR), and UDP-glucuronosyltransferases (UGS) are transcriptionally induced by a low concentration of a variety of chemical compounds including phenolics, flavonoids, isothiocyanates, and 1,2-dithiole-3-thiones. Induction of the expression of QR and GST subunit genes by phenolic antioxidants has been found to correlate with the chemopreventive properties of phenolic compounds. QR catalyzes the reduction of quinones and protects cells against the cytotoxicity of quinones and their metabolic intermediates.

Isoflavone genistein was shown to be an inducer of human sulfotransferases. Genistein was found to significantly induce mRNA expression of sulfotransferase. Another compound, ethacrynic acid, a plant phenolic acid, is both an inhibitor and an inducer of GST activity. Comparative studies of human colon carcinoma HT29 cells revealed that a number of coordinated factors are involved in ethacrynic acid-enhanced expression of GST and other detoxification enzymes.

Moreover, the ability of rosemary extracts containing phenolic compounds, such as essential oil and its dichloromethane and water-soluble extracts, was studied to modulate the activities of CYP450 and detoxication enzymes. Essential oil was found to selectively induce the activity of CYP2B and enhance the activities of CYP450 and detoxication enzymes (GTS, QR, and UGT).

Butylated hydroxyanisole (BHA) is a synthetic phenolic compound that is a potent inducer of activation enzyme genes. Heme oxygenase 1 (HO-1) is upregulated by detoxification enzyme inducers such as BHA. Research has reported that BHA stimulates HO-1 gene expression in vascular smooth muscle cells, suggesting that BHA represents a potentially novel therapeutic agent in treating or preventing vascular proliferative disease.

The ability of flavonoid compounds to induce the activity of anticarcinogenic enzyme, quinone reductase (QR), has been examined in a wild-type murine hepatoma cell line and in an aryl hydrocarbon receptor-defective mutant of the same cell line. The metabolism of flavonoid inducers including quercetin, tamarixetin, and rhamnetin were demonstrated in both wild-type and mutant cells.

1,2-dithiol-3-thiones, a class of five-membered cyclic sulfur compounds, have chemotherapeutic and chemoprotective properties. 1,2-Dithiol-3-thione and several substituted analogs were found to induce NAD(P)H:quinone reductase activity levels in murine hepatoma cells in culture, thereby enhancing detoxification potential.

Cannabidiols from cannabis plants were studied to induce the expression of CYP1A1 mRNA. The induction of CYP1A 1 expression by cannabidiols was significantly attenuated by the knockdown of the expression of aryl hydrocarbon receptor, a transcription factor that regulates gene expression.

A potential induction mechanism was elucidated by investigating 6-methylsulfinylhexyl isothiocyanate (6-HITC) isolated from broccoli as the GST inducer. 6-HITC was found to activate the antioxidant response element (ARE) and induce nuclear localization of the transcription factor Nrf2 that binds to ARE. The induction by 6-HITC was abolished in the Nrf2-deficient mice, suggesting that 6-HITC is a potential activator of the Nrf2-ARE-dependent detoxification pathway.

16.3 Classification of Metabolic Enzyme Inducers

Two classes of anticarcinogenic enzyme inducers have been proposed: (a) monofunctional and (b) bifunctional inducers. Monofunctional inducers (such as phenolics, isothiocyanates, coumarins, thiocarbamates, cinnamates, 1,2-dithiol-3-thiones, and carotenoids) elevate detoxification enzymes (such as glutathione S-transferases, NAD(P)H:quinone reductase, and UDP-glucuronosyl-transferases) without significantly raising the activity of activation enzymes (such as CYP450 and flavor-containing monooxygenase).

Bifunctional inducers, on the other hand, including polycyclic aromatic hydrocarbons, flavonoids, beta-naphthoflavone, and azo dyes, induce both detoxification enzymes and activation enzymes in xenobiotic metabolism. Bifunctional induction was found to be dependent on aryl hydrocarbon receptor function. A direct assay of the activity of quinone reductase in murine hepatoma cells has been utilized to characterize bifunctional and monofunctional inducers.

Quinone reductase carries out two-electron reductions to protect cells against the toxicity of quinones. It can be induced in many tissues in coordination with major detoxification enzymes that protect against reactive intermediates. Measurements of the activity of quinine reductase in murine hepatoma cells and two mutants provide an efficient method for studying the potency and mechanism of detoxification enzyme induction.

A simple, efficient microtiter plate assay has been developed for the direct measurement of quinine reductase basal activity and inducibility in human peripheral blood lymphocytes grown in suspension culture. Quinine reductase is exclusively induced by monofunctional inducers. A monofunctional inducer elevates detoxification enzyme in various tissues without significantly raising activation enzyme such as CYP450 or aryl hydrocarbon hydroxylase.

Table 16.1 presents typical monofunctional and bifunctional inducers.

Table 16.1 Typical monofunctional and bifunctional inducers

Monofunctional inducers	Bifucntional inducers
Phenols	Polycyclic aromatic hydrocarbons
Isothiocyanates	Flavonoids
Coumarins	Azo dyes
Thiocarbamates	
Cinnamates	
1,2-Dithiol-3-thiones	

In spite of many studies supporting the classification of monofunctional and bifunctional inducers, further investigations are needed to ensure that monofunctional inducers modulate exclusively on either detoxification or activation enzymes, but not both enzymes. Review of these compounds listed in the table reveals that neither monofunctional nor bifunctional inducers display common characteristics in terms of chemical structures or functional groups. Further examination reveals few common properties among metabolic enzyme inducers, except their ability to modify sulfhydryl groups by oxidation, reduction, or alkylation.

16.4 Inducer-Enzyme Interactions

By interacting with metabolic enzyme covalently or non-covalently, an inducer molecule can affect the activity of enzyme by the following two manners: (a) modifying the conformation of the enzyme or (b) affecting reactive intermediate generation via inducer-metabolic enzyme interaction. As a result, the activity of enzyme is either increased or decreased, depending on whether the inducer leads to more or less favorable enzyme conformation for catalytic reactions during xenobiotic metabolism.

Elucidation of a potential change in enzyme conformation or inducer-enzyme interaction is essential to an understanding of underlying factors involving induction or inhibition of metabolic enzymes mediated by an inducer. By studying how inducers affect the interaction of foreign compounds with metabolic enzymes, research has suggested that inducers covalently react with cellular molecules by virtue of highly reactive functional group such as sulfhydryl. Nevertheless, investigations of enzyme conformation with respect to inducer-metabolic enzyme interaction have been lacking.

To exert their toxic effects, most foreign compounds require activation catalyzed by activation enzymes, mostly cytochrome P450, to form electrophilic reactive intermediates or metabolites. The reactive groups of xenobiotic reactive intermediate or metabolite are substrates of metabolic enzymes. They are essentially electrophiles, which are either positively charged or have an atom carrying a partial positive charge.

An electrophilic reactive intermediate or metabolite prefers to interact with a nucleophilic group in the amino acid side chain of metabolic enzymes. Potentially nucleophilic groups in the amino acid side chains of metabolic enzymes are serine and tyrosine hydroxyl, aspartate or glutamate carboxylase, histidine imidazole, and cysteine sulfhydryl groups.

The substrate for detoxification enzyme conjugation reaction often shares a common feature that contains an electrophilic atom. In detoxification enzyme-catalyzed conjugation reaction, the site of glucuronidation is generally an electron-rich nucleophilic O, N, or S atom. Glucuronidation is the primary metabolic reaction for many compounds containing functional groups such as -OH, -COOH, -SH, and -NH2.

16.4.1 Michael Reaction Acceptors

Induction of detoxification enzymes is a major strategy for reducing the susceptibility of living cells to the toxic and carcinogenic effects caused by metabolic reactive intermediates including reactive oxygen species and free radicals. A diversity of small molecules of naturally occurring or synthetic origins are capable of inducing detoxication enzymes. Evaluation of their chemical structures reveals that detoxification enzyme inducers belong to a variety of chemical classes that have the ability to modify sulfhydryl group of a cysteine residue in metabolic enzymes.

Moreover, extensive research in inducer-enzyme interactions has revealed that many detoxication enzyme inducers contain Michael acceptor functionalities, such as in olefins or acetylenes. The potency of inducers was also found to parallel with their reactivity as Michael acceptors. Olefins (alkenes), unsaturated hydrocarbons, are made up of hydrogen and carbon that contain one or more pairs of carbon-carbon double bond (C=C). Acetylenes are chemical compounds that consist of the formula of HC≡CH (carbon-carbon triple bond). Michael acceptors are capable of conjugation with electron-withdrawing groups, such as carbonyl [C=O], sulfinyl [S=O], nitrile [C≡N], or acetylene [C≡C].

Table 16.2 lists detoxification enzyme inducers with Michael acceptor functionalities.

Typical functional groups of Michael reaction receptors are alpha, beta-unsaturated aldehydes, ketones, quinones, thioketones, sulfones, esters, nitriles, and nitro groups.

16.4.2 Unsaturated Carbon-Carbon Bonds

Many of the inducers contain Michael reaction acceptors with functionalities such as olefins or acetylenes. Olefins are unsaturated hydrocarbons that contain one or more pairs of carbon atoms linked by a double bond. Acetylenes are also unsaturated

Table 16.2 Typical metabolic enzyme inducers with Michael reaction acceptors

Classes	Compounds
Sulforaphane and isothiocycanate	Sulforaphane
	Phenylethyl isothiocyanate
1,2-Dithiole-3-thione and derivatives	1,2-Dithioe-3-thione (D3T)
	4-Methyl-5-pyrazinyl-D3T (OPZ)
Indole-3-carbinol	Indole-3-carbinol
Flavonoids	Catechin
Epicatechin	Epigallocatechin
	Leucocyanidines
Quercetin	Myricetin
	Fisetin
Isoliquiritigenin	Diosmin
Hesperidin	
Isoflavones	Genistein
Daidzein	
Phenols and polyphenols	Resveratrol
	Curcumin
Gallic acid	
Rosmarinic acid	
Carnosic acid	
Ellagic acid	Tannin
Protocatechuic acid	
Organosulfur	Diallyl sulfide
Diallyl disulfide	
Diallyl trisulfide	
Alliin	
Terpenes and terpenoids	Beta-carotene
	Lycopene
Canthaxanthin	
Astaxanthin	
Zerumbone	
Limonene	
Quinoline	Ethoxyquinn
Others	Nivalenol

hydrocarbons but consist of two carbon atoms bonded together by a triple bond. The carbon-carbon double or triple bond serves as a source of electrons, whose availability is determined by the groups attached to it. An electron-withdrawing group attached to the carbon-carbon double or triple bond (such as C=O, -COOH, -COOR, or -CN) is capable of activating the double or triple bond toward the reagents that are electron-rich.

16.4.3 *Phenolic Hydroxyl Groups*

Studies of metabolic enzyme inducers revealed that the presence of hydroxyl groups at ortho-position on the aromatic rings is an important structural element for high inducer potency.

Ortho-hydroxyl groups have a significant impact on the protective role of detoxification enzyme inducers. Table 16.3 presents typical inducers that contain an ortho-hydroxyl group on the aromatic ring.

Inducers such as flavonoid and curcuminoid analogs contain phenolic hydroxyl groups in addition to Michael acceptor functionalities. Since phenol hydroxyl groups can scavenge oxygen- and nitrogen-centered reactive intermediates, flavonoids and curcuminoids play not only an indirect protective role by inducing detoxification enzymes but also a direct protective role by scavenging hazardous oxidants.

Introduction of ortho-hydroxyl groups to the aromatic rings of phenylpropenoids was found to dramatically enhance their potencies not only as inducers of quinone reductase but also as quenchers of superoxide. The presence of an ortho-hydroxyl substituent on the aromatic ring was also found to profoundly increase the induction potency of benzylidene-alkanones on quinone reductase activity, which also appears to be correlated with their ability to quench superoxide radicals.

16.5 Role of Antioxidant Response Element

Many detoxification enzyme genes are regulated by antioxidant-responsive elements (ARE) that are targets of the transcription factor E2-related factor 2 (Nrf2). Experiments have demonstrated that when Nrf2 is deleted in knockout mice, the basal levels of detoxification enzymes are very low and are not inducible, resulting in these mice being much more susceptible to benzo[a]pyrene carcinogenesis than their wild-type counterparts.

Table 16.3 Metabolic enzymes inducers with ortho-hydroxyl group on aromatic ring

Class of compound	Chemical modulator
Flavonoids	Catechin
Epicatechin	Epigallocatechin
	Leucocyanidines
Quercetin	Myricetin
	Fisetin
Hesperidin	Diosmin
Phenols and polyphenols	Curcumin
Gallic acid	
Rosmarinic acid	
Carnosic acid	
Ellagic acid	Tannin
Protocatechuic acid	

It has been proposed that induction of detoxification enzymes that protect cells against cytotoxicity and carcinogenesis of electrophilic reactive intermediates is mediated by activation of ARE. Inducers of detoxification enzymes were found to disrupt the cytoplasm complex between the protein Keap1 and the transcription factor Nrf2, thereby releasing Nrf2. Nrf2 then migrates to the nucleus where it activates the ARE of detoxification genes and accelerates their transcription. Consequently, the induction of detoxification enzymes requires the functional integrity of both the repressor Keap1 and the transcription factor Nrf2.

The reaction of inducer with Keap1 results in the formation of intermolecular disulfide bridges. Inducer disrupts the Keap1-Nrf2 complex by modifying cysteine residues of Keap1, thereby releasing Nrf2 to migrate to the nucleus, where inducer activates ARE and accelerates their transcription. The Keap1-Nrf2-ARE pathway is proposed to control a network of cytoprotective genes that defend against oxidative stress and inflammation.

Bibliography

Albena T, Dinkova-Kostova MA, Massiah RE, Bozak RJ, Talalay P (2001) Potency of Michael reaction acceptors as inducers of enzymes that protect against carcinogenesis depends on their reactivity with sulfhydryl groups. Proc Natl Acad Sci 98(6):3404–3409

Chen CH (2012) Activation and detoxification enzymes: functions and implications. Springer Sciences, New York

Chen Y, Huang C, Zhou T, Chen G (2008) Genistein induction of human sulfotransferases in HepG2 and Caco-2 cells. Basic Clin Pharmacol Toxicol 103(6):553–559

De Long MJ, Dolan P, Santamaria AB, Bueding E (1986) 1,2-Dithiol-3-thione analogs: effects on NAD(P)H:quinone reductase and glutathione levels in murine hepatoma cells. Carcinogenesis 7 (6):977–980

De Long MJ, Santamaria AB, Talalay P (1987) Role of cytochrome P1-450 in the induction of NAD(P)H:quinone reductase in a murine hepatoma cell line and its mutants. Carcinogenesis 8 (10):1549–1553

Debersac P, Heydel JM, Amiot MJ, Goudonnet H, Artur Y, Suschetet M, Siess MH (2001) Induction of cytochrome P450 and/or detoxication enzymes by various extracts of rosemary: description of specific patterns. Food Chem Toxicol 39(9):907–918

Dinkova-Kostova AT, Holtzclaw WD, Cole RN, Itoh K, Wakabayashi N, Katoh Y, Yamamoto M, Talalay P (2002) Direct evidence that sulfhydryl groups of Keap1 are the sensors regulating induction of phase 2 enzymes that protect against carcinogens and oxidants. Proc Natl Acad Sci 99(18):11908–11913

Dinkova-Kostova AT, Fahey JW, Talalay P (2004) Chemical structures of inducers of nicotinamide quinone oxidoreductase 1 (NQO1). Methods Enzymol 382:423–448

Fong AT, Swanson HI, Dashwood RH et al (1990) Mechanisms of anti-carcinogenesis by indole-3-carbinol. Studies of enzyme induction, electrophile-scavenging, and inhibition of aflatoxin B1 activation. Biochem Pharmacol 39:19–26

Jaiswal AK, Venugopal R, Mucha J, Carothers AM, Grunberger D (1997) Caffeic acid phenethyl ester stimulates human antioxidant response element-mediated expression of the NAD(P)H: quinone oxidoreductase (NQO1) gene. Cancer Res 57(3):440–446

Liu XM, Azam MA, Peyton KJ, Ensenat D, Keswani AN, Wang H, Durante W (2007) Butylated hydroxyanisole stimulates heme oxygenase-1 gene expression and inhibits neointima formation in rat arteries. Cardiovasc Res 74(1):169–179

Lnenickova K, Skalova L, Stuchlikova L, Szotakova B, Matouskova P (2018) Induction of xenobiotic-metabolizing enzymes in hepatocytes by beta-naphthoflavone: time-dependent changes in activities, protein and mRNA levels. Acta Pharm 68(1):75–85

Morimitsu Y, Nakagawa Y, Hayashi K, Fujii H, Kumagai T, Nakamura Y, Osawa T, Horio F, Itoh K, Iida K, Yamamoto M, Uchida K (2002) A sulforaphane analogue that potently activates the Nrf2-dependent detoxification pathway. J Biol Chem 277(5):3456–3463

Prestera T, Talalay P (1995) Electrophile and antioxidant regulation of enzymes that detoxify carcinogens. Proc Natl Acad Sci 92(19):8965–8969

Prochaska HJ, Talalay P (1988) Regulatory mechanisms of monofunctional and bifunctional anticarcinogenic enzyme inducers in murine liver. Cancer Res 48(17):4776–4782

Prochaska HJ, De Long MJ, Talalay P (1985) On the mechanisms of induction of cancer-protective enzymes: a unifying proposal. Proc Natl Acad Sci 82(23):8232–8236

Pugazhenthi S, Akhov L, Selvaraj G et al (2007) Regulation of heme oxygenase-1 expression by demethoxy curcuminoids through Nrf2 by a PI3-kinase/Akt-mediated pathway in mouse beta-cells. Am J Physiol Endocrinol Metab 293:E645–E655

Ramos-Gomez M, Kwak MK et al (2001) Sensitivity to carcinogenesis is increased and chemoprotective efficacy of enzyme inducers is lost in nrf2 transcription factor-deficient mice. Proc Natl Acad Sci U S A 98:3410–3415

Rushmore TH, Pickett CB (1990) Transcriptional regulation of the rat glutathione S-transferase Ya subunit gene. Characterization of a xenobiotic-responsive element controlling inducible expression by phenolic antioxidants. J Biol Chem 265(24):14648–14653

Shen H, Ranganathan S, Kuzmich S, Tew KD (1995) Influence of ethacrynic acid on glutathione S-transferase pi transcript and protein half-lives in human colon cancer cells. Biochem Pharmacol 50(8):1233–1238

Talalay P (1989) Mechanisms of induction of enzymes that protect against chemical carcinogenesis. Adv Enzyme Regul 28:237–250

Talalay P, Fahey JW, Holtzclaw WD, Prestera T, Zhang Y (1995) Chemoprotection against cancer by phase 2 enzyme induction. Toxicol Lett 82–83:173–179

Ushida Y, Talalay P (2013) Sulforaphane accelerates acetaldehyde metabolism by inducing aldehyde dehydrogenases: relevance to ethanol intolerance. Alcohol Alcohol 48(5):526–534

Yamaori S, Kinugasa Y, Jiang R, Takeda S, Yamamoto I, Watanabe K (2015) Cannabidiol induces expression of human cytochrome P450 1A1 that is possibly mediated through aryl hydrocarbon receptor signaling in HepG2 cells. Life Sci 136:87–93

Yang CS, Chhabra SK, Hong JY et al (2001) Mechanisms of inhibition of chemical toxicity and carcinogenesis by diallyl sulfide (DAS) and related compounds from garlic. J Nutr 131:1041S–1045S

Yannai S, Day AJ, Williamson G, Rhodes MJ (1998) Characterization of flavonoids as monofunctional or bifunctional inducers of quinone reductase in murine hepatoma cell lines. Food Chem Toxicol 36(8):623–630

Chapter 17
Induction of Metabolic Enzymes for Health Effects

Oxidative stress occurs when xenobiotic metabolic reactive intermediates, including reactive oxygen species or free radicals, are neither readily detoxified by detoxification enzymes nor compensated for by the body's antioxidant defense systems. Oxidative stress is associated with inflammation and damages to cells and tissues. The ability of individuals to remove chemically reactive metabolites from the body plays a crucial role in the reduction of the risk of a variety of disease conditions.

The mechanisms underlying the defense against xenobiotics-mediated toxicity include (a) a reduction of metabolic activation by appropriately inhibiting activation enzymes, (b) an enhancement of metabolic detoxification by inducing detoxification enzymes, and (c) an increase of antioxidant defense. In conjunction with (a) and (b), an important feature of xenobiotic metabolic enzymes is their ability to be activated or inhibited by modulation compounds.

Modulators are small molecules that exhibit induction or inhibition effects on the activities of activation enzymes or detoxification enzymes. Meanwhile, detoxification systems exhibit a great deal of individual variability and are affected by genetics, lifestyle, and environmental factors.

The action of a small modulation molecule can be direct by binding to an enzyme, either covalently or non-covalently, thereby affecting the structure or conformation of the enzyme in a manner that either increases or decreases its enzymatic activity. To exert such action on the enzyme, it is essential that the small modulator is able to transport across cell membranes.

There is convincing evidence that consumption of certain dietary ingredients may favorably modulate biotransformation of xenobiotic toxicities. The available information suggest that people whose diets are rich in vegetables and fruits have a lower risk of developing disease conditions. Particularly, diets rich in inducers of detoxification enzyme have been found to promote the scavenge of oxidants, resulting in a positive influence on health and related aging processes.

Inducers of detoxification enzymes act as indirect antioxidants by boosting the body's antioxidant systems, thus exerting antioxidant activities. Metabolic enzymes that act catalytically to detoxify reactive intermediates, including reactive oxygen

C.-H. Chen, *Xenobiotic Metabolic Enzymes: Bioactivation and Antioxidant Defense*,
https://doi.org/10.1007/978-3-030-41679-9_17

species and free radicals, are regulated by transcription factor Nrf2. The Keap1-Nrf2-ARE pathway controls a network of genes, which defends against the damaging effects of reactive electrophile-mediated oxidative stress and its associated inflammation and disease conditions.

Unlike inducers of metabolic enzymes, antioxidant molecules, such as glutathione and vitamins C and E, serve directly in antioxidant defense by scavenging free radicals or other oxidant molecules. It has been postulated that the ingestion of an excessive amount of antioxidant is presumed to shift the oxidant-antioxidant balance toward the antioxidant side.

Although antioxidants may effectively break down reactive oxygen species and free radicals, such a shift of balance toward the antioxidant side by taking excessive antioxidants could result in affecting some key physiological processes that are dependent on free radicals. This issue is a matter of concern and requires further investigation.

17.1 Metabolic Enzyme Modulation

An important toxicologically relevant feature associated with xenobiotic metabolism is the inducibility of metabolic enzymes. While some xenobiotics exhibit intrinsic toxicity, some others are metabolically activated to potential toxicants by activation enzymes. The generated toxic metabolites then undergo detoxification enzyme-catalyzed reactions before they are ready for excretion from the body. Consequently, an induction or inhibition of these metabolic enzymes has a significant impact on the extent of toxicity of xenobiotics.

Enzyme inducibility is of major significance because it plays a role in the susceptibility of individuals to xenobiotics-mediated toxicities. Strategies for protecting cells from initiation toxic events include (a) a decrease in the expression of activation enzymes responsible for generating reactive intermediate species and (b) an increase in the activities of detoxification enzymes that detoxify electrophilic intermediates and free radicals known to intervene in regular cellular processes.

Nevertheless, in either of these two strategies, it is essential to realize that the consequence of enzyme induction or inhibition is dependent on not only the specific effect on activation or detoxification enzymes but also the balance between the activities of activation and detoxification enzymes. Furthermore, the expression of metabolic enzymes can also differ significantly as a result of diet, lifestyle, and exposure to environmental chemicals.

Induction of activation enzymes leading to a higher expression of enzyme activity can result in an enhancement in the rate of activation and the production of potentially toxic metabolic intermediates. If the generated reactive intermediates are not quickly removed from the body, the induction of activation enzymes would promote toxic effects of foreign compounds. On the other hand, the induction of detoxification enzymes can lead to a higher expression of enzyme activity and an increase in the rate of detoxifying reactions, leading to a decrease in xenobiotic toxic effect and an acceleration of excretion of foreign compounds from the body.

Moreover, the effects of modulating activation enzymes or detoxification enzymes on xenobiotic toxicity depend on not only the relative activities of these enzymes but also the nature of the foreign compounds. For a foreign compound whose toxicity is caused by metabolic activation, inhibition of activation enzymes would decrease toxic effects derived from xenobiotic metabolism. While for a foreign compound with an inherent toxicity, the induction of detoxification enzymes would decrease xenobiotic toxicity. A more detailed analysis of the effects of metabolic enzyme modulation is presented below.

17.1.1 Modulation of Activation Enzymes

Figure 17.1 illustrates how induction and inhibition of activation enzyme may potentially affect xenobiotic toxicity. When a foreign compound is metabolized to produce a potent reactive intermediate or metabolite, a significant induction of activation enzyme to a higher activity can increase the rate of its production, resulting in an enhancement in xenobiotic toxic effect.

The figure also reveals that a significant inhibition of activation enzyme may also enhance xenobiotic toxic effect, due to an increase in an impairment in metabolic clearance of foreign compound. However, modest inhibition of activation enzyme may result in a lower xenobiotic toxic effect, assuming that the resulted enzyme activity is still enough to perform its metabolic function.

Among activation enzymes, CYP450 is most studied with respect to enzyme modulation.

A modest inhibition of CYP450 leads to a lower activity, which modestly decreases the activation rate and produces less reactive intermediate. As a result, a reduction in xenobiotic toxic effect may occur. However, an over-inhibition of activation enzyme leads to an accumulation of foreign compounds, which can cause impaired metabolic clearance.

A typical example of inducer of activation enzyme is phenobarbital. Investigations revealed that phenobarbital induces drug metabolism in animal models and humans. The induction effect of phenobarbital mainly involves the activation of CYP450 enzymes. Phenobarbital was found to enhance the activation of cocaine, resulting in the potentiation of cocaine-induced hepatotoxicity.

In contrast, grapefruit juice components (naringin and furanocoumarins) are typical examples of activation enzyme inhibitor. The consumption of grapefruit juice with drugs taken orally has been reported to inhibit the activity of intestinal CYP3A4, which results in a decrease in the metabolism of many drugs.

It should point out that the induction or inhibition of activation enzyme, leading to increasing or decreasing the toxicity of a foreign compound, is an issue of complexity. The effects of activation enzyme modulation are dependent not only on the

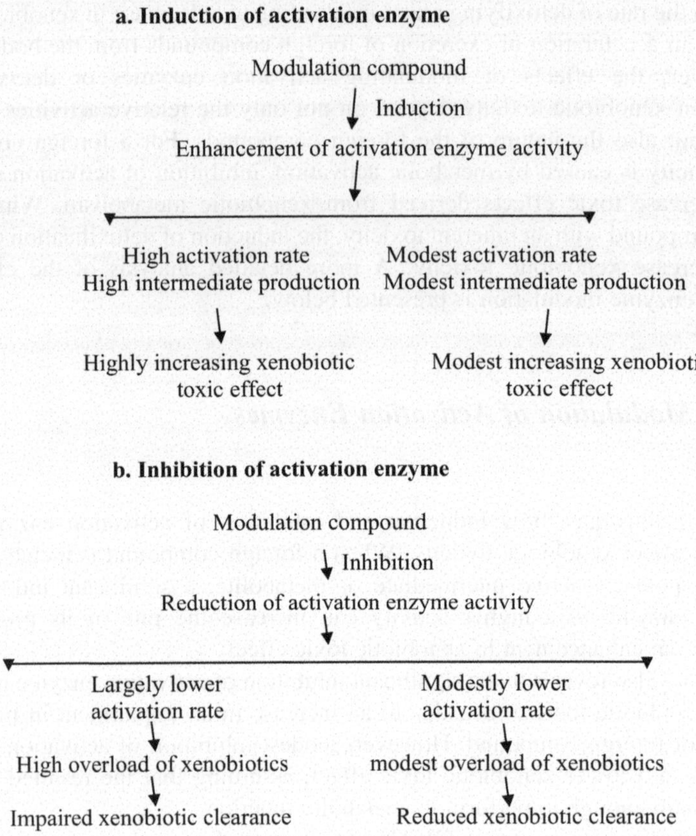

Fig. 17.1 Activation enzyme modulation affecting xenobiotic metabolism. (**a**) Induction of activation enzyme. (**b**) Inhibition of activation enzyme

resulted activity of activation enzyme but also on the rate of respective detoxification enzyme.

17.1.2 Modulation of Detoxification Enzymes

Generally, induction of detoxification enzymes enhances the rate of detoxification reactions, leading to a faster activity in forming conjugated metabolites and resulting in a decrease in xenobiotic toxic effect, thus facilitating the excretion of xenobiotics. Conversely, inhibition of detoxification enzymes decreases the rate of detoxification reactions, leading to a lower reaction in forming conjugated metabolites and resulting in an enhancement of xenobiotic toxic effect. In other words, the induction

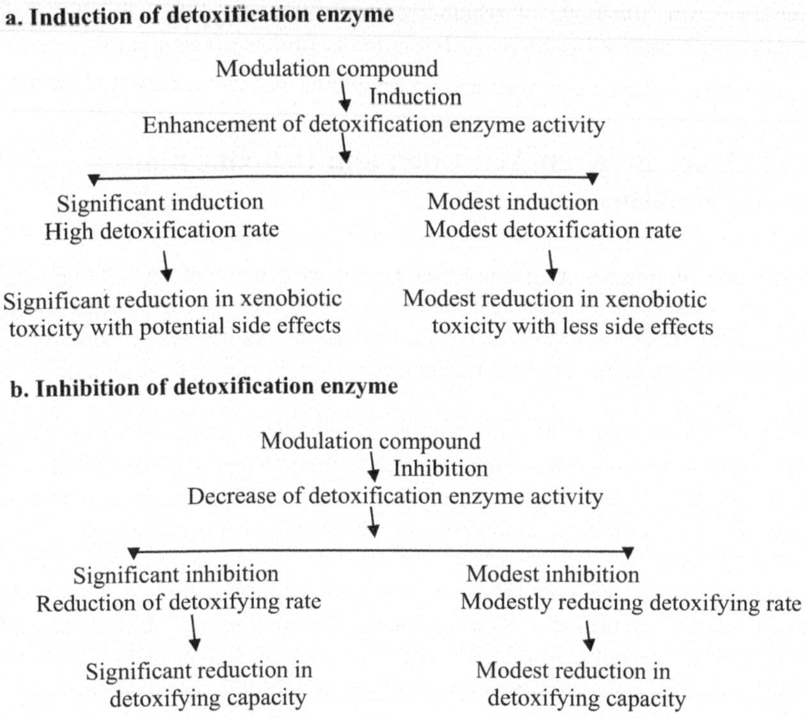

Fig. 17.2 Detoxification enzyme modulation affecting xenobiotic metabolism. (**a**) Induction of detoxification enzyme. (**b**) Inhibition of detoxification enzyme

or inhibition of detoxification enzyme can result in a lower or higher xenobiotic toxic effect.

Among detoxification enzymes, UDP-glucuronosyl transferase (UGT), glutathione S-transferases (GST), and NAD(P)H:quinone reductase (NQO) are most studied with respect to enzyme modulation. A typical detoxification enzyme inducer is oltipraz, a 1,2-dithiole-3-thione derivative, which was reported to cause several-fold inductions of constitutive hepatic and gastric activities of GST and NQO in mice.

An increase in the expression of detoxification enzyme genes by oltipraz was also found to attribute to a reduction in aflatoxin B-induced hepatocarcinogenesis. Moreover, phenobarbital was reported to induce glucuronide conjugation of 4-hydroxyl-NNK catalyzed by UGT. NNK (4-(Methylnitrosamino)-1-(3-pyridyl)-1butanone) is a tobacco-specific carcinogen, which exerts carcinogenic potential through alpha-hydroxylation in metabolic activation.

Figure 17.2 illustrates how induction and inhibition of detoxification enzymes affect xenobiotic toxic effect. Comparison of Fig. 17.2 with Fig. 17.1 reveals that the modulation of detoxification enzyme is less complicated than that of activation enzyme. However, the issue of whether complication may arise if detoxification enzyme is overly induced is an issue that requires further investigation.

17.2 Balance Between Activation and Detoxification Metabolisms

The efficiency of foreign compound metabolism is dependent on relative rates or relative activities of activation and detoxification enzymes. In general, minimal xenobiotic toxicity is found in either (a) a fine balance rate between activation and detoxification expressions or (b) a higher expression of detoxification activity than activation activity.

Figure 17.3 summarizes the relative rates of activation and detoxification expressions in relation to xenobiotics-mediated toxicities. When the rate of activation expression is higher than that of detoxification expression, reactive intermediates generated from xenobiotic metabolism are overload, which can result in higher xenobiotic toxicity.

In contrast, when the rates of activation and detoxification are balanced, the generated reactive intermediates are present at a minimum, which can lead to lower xenobiotic toxicity. Similarly, when the rate of activation is significantly lower than the rate of detoxification, reactive xenobiotic intermediates are also present at a minimum, which may also result in a lower xenobiotic toxicity. However, it should point out that when the rate of detoxification metabolism is overly induced, the question that whether complication may arise is an issue of concern. Further investigations to address this concern are needed.

The following example is given to elaborate the above figure. Induction of CYP450 expression is likely to either increase or decrease the toxic effects of xenobiotics, depending on the relative activity of detoxification enzymes. When the induction of CYP450 results in the rate of activation comparable with that of detoxication, such induction is likely to produce beneficial effects. Nevertheless, if the induction of CYP450 results in the rate of activation being significantly higher

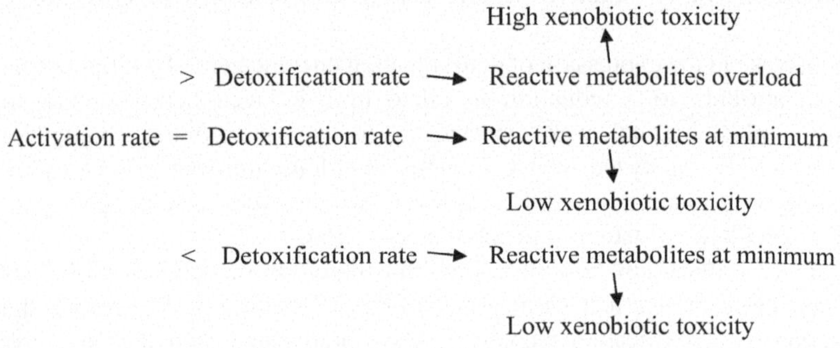

Fig. 17.3 Relative rates of metabolic enzymes affecting xenobiotic toxicity

than that of detoxification, then the induction may produce harmful effects, such as in the case of smoking-related induction of CYP450.

Consequently, a fine balance between the expressions of activation enzyme and detoxification enzyme determines whether reactive intermediates such as reactive oxygen species are adequately detoxified or may cause toxic effects. In general, the rate of reactive intermediate production comparable with that of detoxification reaction is essential to achieve this fine balance. Accordingly, detoxification metabolism that functions appropriate appears to be an important means of preventing xenobiotics-mediated toxic effects.

17.3 Enzyme Modulation as Defense Against Potential Toxic Effects

Foreign compound-mediated toxic effects occur at least in the following three circumstances: (1) xenobiotics are inherent toxicants which require detoxification by antioxidants and detoxification enzymes; (2) xenobiotics exhibit toxicity due to the formation of reactive intermediates or metabolites generated by metabolic activation, but are not effectively detoxified by antioxidants and detoxification enzymes; and (3) xenobiotic toxic effects are potentiated by environmental chemicals or lifestyle (smoke or alcohol) that are capable of interfering effectively with detoxification enzymes.

It has been proposed that modulation of foreign compound-metabolizing enzymes may be a useful approach in minimizing the risk of xenobiotics-mediated toxic effects. One proposed hypothesis is to inhibit the expression of activation enzymes as well as induce that of detoxification enzymes. Another hypothesis proposes that the induction of detoxification enzymes alone is sufficient to achieve the protection. In general, the inhibition of activation enzymes to affect xenobiotic toxicities is more complicated than the induction of detoxification enzymes. Extensive studies have been carried out to test the latter hypothesis.

Results of these investigations suggested that the induction of detoxification enzymes appears to be an effective means for achieving protection against a variety of toxicities and carcinogens. In line with these studies, it has been proposed that a promising approach to minimize xenobiotic toxic effects and related oxidative stress is to increase the intake of dietary inducers of detoxification enzymes.

17.4 Strategies Against Xenobiotic Toxicity

As discussed above, the susceptibility of organisms to toxic effects of reactive intermediates or metabolites depends on the balance in the efficacy between activation enzymes and detoxification enzymes. Extensive research has been focused on the modulation of metabolic enzymes to overcome the imbalance between their activities.

Foreign compounds are initially metabolized by activation enzymes, mainly CYP450s, to generate either inactive metabolites or reactive intermediates. Some researchers have focused on the inhibition of activation enzymes to keep the production of reactive intermediates low enough, so that even those with less effective detoxification enzymes the body still can deal with these potential toxicants. However, inhibition of activation enzyme such as CYP450s to a level of activity lower than that of metabolic detoxification can hinder metabolic activation of certain toxins or procarcinogens, leading to ineffectiveness in removing these toxic substances from the body.

Some researchers have worked on the approach to selectively modulate the expression of detoxification enzymes, such as UDP-glucuronosyl transferases, glutathione S-transferases, and quinone reductase, to make the body better able to deal with reactive intermediates which it encounters. Accordingly, it has been proposed that induction of detoxification enzymes is a major potential strategy for reducing the risk of oxidative stress-associated disease conditions such as inflammation, degenerative diseases, and cancers.

Animal studies have supported the proposal that induction of detoxification enzymes is a sufficient condition for obtaining chemoprevention. For example, investigations in rodent models for chemoprevention of oltipraz (a strong detoxification enzyme inducer) on aflatoxin-induced hepatocarcinogenesis have demonstrated that increased expression of detoxification gene is of crucial importance. Furthermore, selective induction of detoxification enzymes seems to offer a safer prospect to achieve protection against toxic and carcinogenic effects of foreign compounds.

Researchers have actively searched for monofunctional inducers of detoxification enzymes. Nevertheless, among reported monofunctional inducers of detoxification enzymes, many of them may also inhibit or induce CYP450 isozymes. Moreover, it is important to keep in mind that a potential problem may occur, if a modulator alters the action of therapeutic drugs due to inducer-drug interactions.

17.5 Roles of Nrf2 and ARE on Enzyme Modulation

The nuclear transcription factor E2-related factor 2 (Nrf2) has been shown to play crucial roles in the mechanisms of preventing xenobiotics-generated toxicity and carcinogenesis. The protective roles of Nrf2 were found to attribute in part to its involvement in the induction of detoxification enzymes. Experiments have demonstrated that when Nrf2 is deleted in knockout mice, the basal levels of detoxification enzymes are very low and are not inducible. Consequently, knockout mice are much more susceptible to benzo[a]pyrene carcinogenesis than their wild-type counterparts. Moreover, constitutive hepatic and gastric activities of glutathione S-transferases and quinone oxidoreductase were reduced to less than half in Nrf2-deficient mice compared with wild-type counterparts.

As discussed in a previous chapter, transcriptional control of the expression of detoxification enzymes is mediated in part through the antioxidant response element (ARE) found in the regulatory regions of detoxification genes. Binding of the transcription factor Nrf2 to the ARE appears to be essential for the induction of detoxification enzymes. Research has shown that exposure of rodents to 1,2-dithiole-3-thione (a detoxification enzyme inducer) triggers nuclear accumulation of the transcription factor Nrf2 and enhances its binding to ARE, leading to transcriptional activation of genes involved in detoxifying carcinogens.

17.6 Lifestyle Modulation of Metabolic Enzymes

In addition to the above-discussed inhibition and induction of metabolic enzymes, lifestyle can also modulate the expressions of xenobiotic metabolic enzymes. Two most known lifestyle factors that are capable of modulating foreign compound-metabolizing enzymes are alcohol and cigarette smoke. The effects of alcohol and cigarette smoke on the expression of metabolic enzymes are described below.

17.6.1 Alcohol

The metabolisms of foreign compounds are affected by chronic alcohol consumption. CYP isoform plays a key role in alcoholic-associated oxidative stress and liver disease. Enzyme CYP2E1 is the primary component of alcohol oxidizing system in the hepatic microsomes. CYP2E1-catalyzed reaction is involved in the production of reactive oxygen species. Induction of CYP2E1 by alcohol increases metabolic activation of carcinogens. Accordingly, an enhanced vulnerability of the alcoholic to the hepatotoxicity of many foreign compounds has been recognized. Moreover, heavy drinkers develop severe hepatotoxicity from chemicals such as carbon tetrachloride exposure, but not for non-drinkers.

17.6.2 Cigarette Smoke

Activation enzymes induced by cigarette smoking may increase the risk of cancer by enhancing the metabolic activation of carcinogens. Benzo[a]pyrene and nitrosamines in cigarette smoke undergo metabolism by activation enzymes, mainly CYP450. Polycyclic aromatic hydrocarbons in cigarette are believed to be responsible for inducing CYP450s, such as CYPlAl, CYP1A2, and perhaps CYP2El. The induction of CYP450 activity by cigarette smoke also plays a major role in smoke-drug interactions.

Bibliography

Buetler TM, Gallagher EP, Wang C (1995) Induction of phase I and phase II drug metabolizing enzyme mRNA, protein, and activity by BHA, ethoxyquin, and oltipraz. Toxicol Appl Pharmacol 135:45–57

Chen CH (2012) Activation and detoxification enzymes: functions and implications. Springer Sciences, New York

Conney AH (2003) Enzyme induction and dietary chemicals as approaches to cancer chemoprevention: the Seventh DeWitt S. Goodman Lecture. Cancer Res 63:7005–7031

Cuendet M, Oteham CP, Moon RC et al (2006) Quinone reductase induction as a biomarker for cancer chemoprevention. J Nat Prod 69:460–463

Dinkova-Kostova AT, Talalay P (2008) Direct and indirect antioxidant properties of inducers of cytoprotective proteins. Mol Nutr Food Res 52(Suppl l):S128–S138

Finley JW (2003) The antioxidant responsive element (ARE) may explain the protective effects of cruciferous vegetables on cancer. Nutr Rev 61:250–254

Hodges RE, Minich DM (2015) Modulation of metabolic detoxification pathways using foods and food-derived components: a scientific review with clinical application. J Nutr Metab 2015:760689

Jakel RJ, Townsend JA, Kraft AD, Johnson JA (2007) Nrf2-mediated protection against 6-hydroxydopamine. Brain Res 1144:192–201

Kensler TW, Curphey TJ, Maxiutenko Y et al (2000) Chemoprotection by organosulfur inducer of phase 2 enzymes: dithiolethiones and dithiins. Drug Metabol Drug Interact 17(1–4):3–22

Liu Y, Kern JT, Walker JR et al (2007) A genomic screen for activators of the antioxidant response element. Proc Natl Acad Sci 104:5205–5210

Manson MM, Ball HW, Barrett MC et al (1997) Mechanism of action of dietary chemoprotective agents in rat liver: induction of phase I and II drug metabolizing enzymes and aflatoxin B I metabolism. Carcinogenesis 18:1729–1738

Murphy SE, Nunes MG, Hatala MA (1997) Effects of phenobarbital and 3-methylcholanthrene induction on the formation of three glucuronide metabolites of 4-(metIlylnitrosamino)-I-(3pyridyl)-I-butanone, NNK. Chem Biol Interact 103:153–166

Murray M (2006) Altered CYP expression and function in response to dietary factors: potential roles in disease pathogenesis. Curr Drug Metab 7:67–81

Nguyen T, Sherratt PJ, Pickett CB (2002) Regulatory mechanisms controlling gene expression mediated by the antioxidant response element. Annu Rev Pharmacol Toxicol 43:233–260

Okey AB, Roberts EA, Harper PA et al (1986) Induction of drug-metabolizing enzymes: mechanisms and consequences. Clin Biochem 19:132–141

Prochaska HJ, TaJalay P (1998) Regulatory mechanisms ofmonofunctionaJ and bifunctional anticarcinogenic enzyme inducers in murine liver. Cancer Res 48:4776–4782

Prochaska HJ, Santamaria AB, Talalay P (1992) Rapid detection of inducers of enzymes that protect against carcinogens. Proc Natl Acad Sci 89:2394–2398

Ramos-Gomez M, Kwak MK, Dolan PM et al (2001) Sensitivity to carcinogenesis is increased and chemoprotective efficacy of enzyme inducers is lost in nrf2 transcription factor-deficient mice. Proc Natl Acad Sci 98:3410–3415

Shen G, Kong AN (2009) Nrf2 plays an important role in coordinated regulation of Phase II drug metabolism enzymes and Phase III drug transporters. Biopharm Drug Dispos 30:345–355

Talalay P (1989) Mechanisms of induction of enzymes that protect against chemical carcinogenesis. Adv Enzyme Regul 28:237–250

Talalay P, Dinkova-Kostova AT, Holtzclaw WD (2003) Importance of phase 2 gene regulation in protection against electrophile and reactive oxygen toxicity and carcinogenesis. Adv Enzyme Regul 43:121–134

Wu L, Noyan Ashraf MH, Facci M et al (2004) Dietary approach to attenuate oxidative stress, hypertension, and inflammation in the cardiovascular system. Proc Natl Acad Sci 101:7094–7099

Zevin S, Benowitz NL (1999) Drug interactions with tobacco smoking. An update. Clin Pharmacokinet 36:425–438

Chapter 18
Diversity of Metabolic Enzyme Modulators

When xenobiotic metabolic reactive intermediates such as reactive oxygen species are involved in cellular toxicity and carcinogenesis, small chemical compounds that modulate the activity of metabolic enzymes could play an important role in the intervention of such cytotoxicity and carcinogenic processes. Accordingly, extensive research has been carried out to investigate chemical compounds that are capable of modulating the activities of activation and detoxification enzymes.

The modulation of activation enzymes is mainly concentrated on cytochrome P450s. CPY450s are responsible for the majority of oxidation reactions involving foreign compound metabolism. The modulation of CYP450s as a defense against xenobiotic-mediated toxic effects is a complex subject. If activation activity is not compensated by detoxification activity, a high expression of CYP450 could increase toxic effects of foreign compounds, due to overload of reactive intermediates or metabolites. In contrast, a low expression of CYP450 could cause an accumulation of reactive intermediates or metabolites, due to impaired metabolic clearance.

Since both activation and detoxification enzymes are involved in the risk of toxicity and carcinogenesis mediated by xenobiotic metabolism, a beneficial enzyme modulator would be an inducer of detoxification enzyme rather than an activator of activation enzyme. Considerable evidence has supported the prospect of modulating the activity of detoxification enzymes, such as UDP-glucuronosyl transferases, glutathione S-transferases, and quinone reductases, as a major strategy to manipulate the risk of toxicity and carcinogenesis mediated by xenobiotic metabolic reactive intermediates and metabolites.

It should be essential to recognize that the balance between the expressions of activation enzymes and detoxification enzymes is also a critical determinant of whether the exposure to toxicants or carcinogens will result in toxic and neoplastic effects or not. In the case of drug metabolism, enzyme induction by enhancing the metabolism of drugs may result in interfering with drug interactions. The rate and extent of drug metabolism may significantly influence the effectiveness of drugs.

An inducer of detoxification enzyme is believed to increase the enzyme expression by binding to the enzyme or activating it so as to enhance the expression of the

C.-H. Chen, *Xenobiotic Metabolic Enzymes: Bioactivation and Antioxidant Defense*,
https://doi.org/10.1007/978-3-030-41679-9_18

gene coding for the enzyme. Extensive research has been carried out to demonstrate that the transcription factor (Nrf2) is essential for the induction of detoxification enzymes. Inducers trigger nuclear accumulation of Nrf2, leading to transcriptional activation of genes involved in carcinogen detoxification. Thus, the increased expressions of detoxifying genes are of critical importance to chemoprevention.

The modulation of detoxification enzymes is mainly focused on the induction of UDP-glucuronosyl transferases, glutathione S-transferases, and quinone reductases. Detoxification enzyme inducers are diverse, including (1) isothiocyanates; (2) 1,2-dithiole-3-thione and its derivatives; (3) indole-3-carbinol; (4) polyphenols, flavonoids, and isoflavones; (5) diallyl sulfides; (6) terpenes, terpenoids, and carotenoids; and (7) other compounds. The modulation of these chemical compounds is briefly discussed below. Some typical chemical structures of enzyme modulators are also presented to show that metabolic enzyme modulators generally consist of functional groups with unsaturated carbon-carbon bonds and hydroxy groups on aromatic rings.

18.1 Isothiocyanates

Isothiocyanates are primarily metabolized through the mercapturic acid pathway, which gives rise to *N*-acetylcysteine conjugates. The effects of isothiocyanates appear to be mediated by the induction of detoxification enzymes, leading to detoxification of electrophilic reactive intermediates or metabolites generated by activation metabolism. It has also been reported that anticarcinogenic effects of isothiocyanates appear to be mediated by the suppression of carcinogen activation catalyzed by CYP450 enzymes through inhibiting CYP450 catalytic activity.

Among isothiocyanate derivatives that have been investigated for their modulation of the activity of metabolic enzymes include sulforaphane and phenethyl isothiocyanate. Sulforaphane (4-methylsulfinylbutane isothiocyanate) is a potent detoxification enzyme inducer. Sulforaphane is thought to provide a protective effect of glucoraphanin. Glucoraphanin, the precursor of sulforaphane, is produced from glucoraphanin by myrosinase. It was reported that sulforaphane significantly induces detoxification enzyme activity in human prostate cells and blocks mammary tumor formation in rats. Oral administration of sulforaphane was also found to potently induce detoxification enzymes in the bladder tissues in animal models.

Phenethyl isothiocyanate was reported to block the metabolic activation of the tobacco-specific nitrosamine, 4-(methylnitrosamino)-1-(3-pyridyl)-1-butanone (NNK), leading to increasing urinary excretion of detoxified metabolites. As a result, phenethyl isothiocyanate is believed to be an effective inhibitor of lung tumor mediated by NNK.

Table 18.1 presents the modulation of isothiocyanate derivatives on xenobiotic metabolic enzymes, including detoxification enzymes and activation enzymes. These findings were derived from investigations in the experiments carried out in animal tissues and cultured cells.

Table 18.1 Modulation of isothiocyanate (ITC) derivatives on xenobiotic metabolic enzymes

Modulation compound	Detoxifying enzyme induction	Activation enzyme inhibition
Sulforaphane	GST, QOR, epoxide hydrolase	CYP2E1, CYP3A4
Allyl-ITC	GST	CYP2C9
Benzyl-ITC	GST	CYP2A6
Phenyl-ITC		CYP2A6
Phenylethyl-ITC	GST, QOR	CYP2A6

Table 18.2 Modulation of 1,2-dithiole-3-thione derivatives on xenobiotic metabolic enzymes

Modulation compound	Detoxifying enzyme induction	Activation enzyme inhibition
1,2-Dithioe-3-thione (D3T)	GST	CYP1A2
Oltipraz	QOR, GST, UGT	CYP1A2
5,6-Dihydrocyclopenta-D3T	QOR, GST	
4-Chloro-5-methyl-D3T	QOR, GST	
4-Phenyl-D3T	QOR, GST	

Structures of sulforaphane and phenylethyl isothiocyanate are shown below:

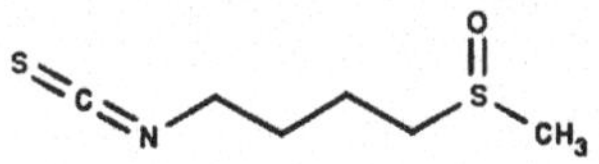

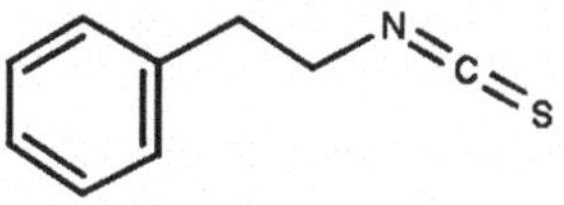

18.2 1,2-Dithiole-3-Thione and Derivatives

1,2-Dithiole-3-thione (D3T), a naturally occurring organosulfur compound, has been found to induce Nrf2-dependent detoxification enzymes. Exposure of rodent to D3T triggers nuclear accumulation of Nrf2, leading to transcriptional activation of genes involved in carcinogen detoxification. Rodent model studies revealed chemoprevention of aflatoxin-induced hepatocarcinogenesis by oltipraz (4-methyl-5-pyrazinyl-3H-1,2-dithiole-3-thione), a synthetic derivative of D3T. The protective effects of D3T and oltipraz are attributed, in part, to the induction of detoxification enzymes such as glutathione S-transferases (GST), UDP-glucuronosyl transferases (UGT), and quinone oxidoreductases (QOR), in addition to the inhibition of activation enzymes, mainly CYP450 enzymes.

Table 18.2 presents various D3T derivatives that are capable of modulating xenobiotic metabolic enzymes, as reported in experiments using animal tissues and cultured cells. The table reveals that oltipraz and other D3T derivatives, 5,6-dihydrocyclopenta-D3T, 4-chloro-5-methyl-D3T, and 4-phenyl-D3T, exhibit the protection against the toxicity of foreign compounds by inducing detoxification enzymes (GST, UGT, or QOR). D3T and oltipraz also inhibit the major activation enzyme, CYP450.

Structures of 1,2-dithiole-3-thione and oltipraz are presented below:

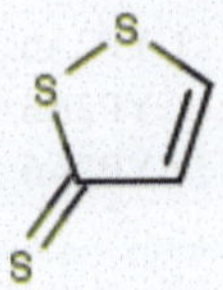

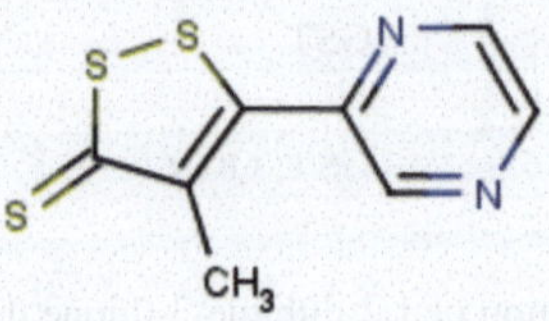

18.3 Indole-3-Carbinol

Similar to isothiocyanate, indole-3-carbinol is a hydrolysis product of glucosinolate. In a number of animal model studies, indole-3-carbinol was reported to induce detoxification enzymes. As an effective chemopreventive agent, indole-3-carbinol was also found to inhibit aflatoxin-induced hepatocarcinogenesis in rats. Table 18.3 presents the modulation of indole-3-carbinol on xenobiotic metabolic enzymes. The table reveals that indole-3-carbinol induces detoxification enzymes (GST and QOR) and inhibits the major activation enzyme, CYP450.

Structure of indole-3-carbinol is shown below:

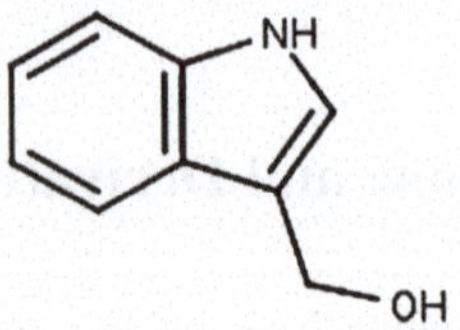

Table 18.3 Modulation of indole-3-carbinol and other compounds on xenobiotic metabolic enzymes

	Detoxification enzyme	Activation enzyme	
Modulation compound	Induction	Inhibition	Induction
Indole-3-carbinol	GST, QOR	CYP1A1	
Geniposide	GTS		
Alliin	GTS, UGT	CYP1A2	CYP2E1

18.4 Polyphenol, Flavonoids, and Isoflavones

Polyphenols, a group of chemical substances found in plants, are characterized by the presence of more than one phenol unit per molecule. There has been a growing interest in the role of phenolic compounds in the prevention of degenerative and cardiovascular diseases and cancers. It has been reported that polyphenols are capable of inducing detoxifying enzymes. In addition, they may also inhibit activation enzymes.

Flavonoids are part of polyphenol class of phytochemicals, such as resveratrol, curcumin, turmeric, carnosol, carnosic acid, protocatechuic acid, tannic acid, and ellagic acid. Isoflavones are related to flavonoids and are mostly found in the bean family, such as genistein, daidzein, and soy isoflavones. Some studies have revealed potential risks from intaking diet high in isoflavones in women with a history of breast cancer. This concern requires further investigations.

18.4.1 Polyphenols

Some naturally occurring polyphenol-containing plants have been reported to inhibit the mutagenicity of chemical carcinogens such as polycyclic aromatic hydrocarbons. Among polyphenols studied, resveratrol is best known for its chemopreventive activity against cytotoxicity and carcinogenesis, suggesting that resveratrol provides protection against oxidative stress-mediated disease conditions. Moreover, animal model studies have revealed that curcumin modestly induces detoxification enzyme activity in the prostate. Induction of detoxification enzymes is a potential mechanism through which the action of curcumin proceeds.

Table 18.4 presents the modulation of some typical polyphenols on xenobiotic metabolic enzymes. These studies were performed in the experiments using animal tissues from rat or mouse or cultured cells from human.

Table 18.4 Modulation of polyphenols on xenobiotic metabolic enzymes

	Detoxification enzymes	Activation enzymes
Modulation compound	Induction	Inhibition
Resveratrol	GTS, QOR, UGT	CYP1A1
Curcumin	GTS, QOR	CYP3A4
Turmeric	GTS	CYP3A4
Carnosol	GTS	CYP1A1
Carnosic acid	GTS	CYP2C9
Protocatechuic acid	GST, UGT, NQO	CYP1A1
Tannic acid	GST, NQO	CYP1A1
Ellagic acid	GST, QR, UGT	CYP1A1
Tea catechins	GST, QR	CYP2E1

Table 18.5 Modulation of flavonoids on xenobiotic metabolic enzymes

	Detoxifying enzyme	Activation enzyme	
Modulation compound	Induction	Induction	Inhibition
Isoliquiritigenin	QOR		
4′-Bromoflavone			CYP1A1
Beta-naphthoflavone	GST, UGT, NQO	CYP2D6	
Leucocyanidines	GST		CYP450
Anthocyanins	GTS		
Catechin	UGT		CYP2E1
Quercetin	QOR, QR		
Myricetin	QOR		

18.4.2 Flavonoids

Flavonoids are the most common group of polyphenolic compounds. A number of in vitro studies have revealed the effects of flavonoids on detoxification and activation enzymes. The protective effects of flavonoids against xenobiotic toxicity are attributed in part to modulating the activities of detoxification enzymes. The modulation of detoxification enzyme (UGT, GST, or QOR) to accelerate detoxifying carcinogens is an important mechanism involving anticarcinogenic effects of flavonoids.

Table 18.5 presents the modulation of typical flavonoids on xenobiotic metabolic enzymes. The table reveals that flavonoids induce detoxification enzyme activity, while the activity of activation enzyme (mainly CYP450) is either induced, inhibited, or unaffected. These studies were carried out in the experiments using cultured cells from human or rat and animal tissues from rat.

Structures of catechin and epicatechin are shown below and so are other flavonoids:

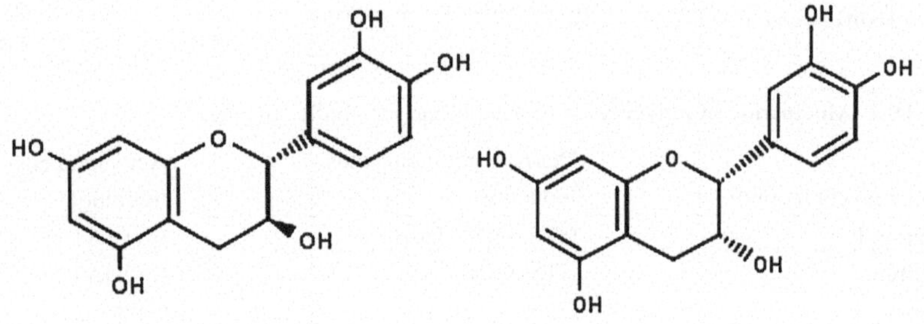

Structures of epigallocatechin and leucocyanidin:

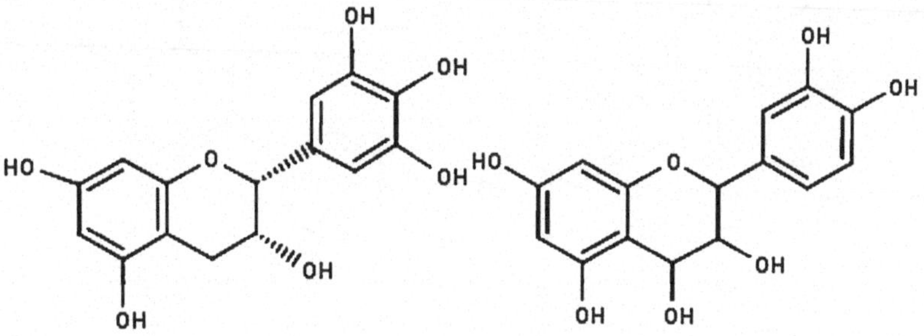

Structures of myricetin and quercetin:

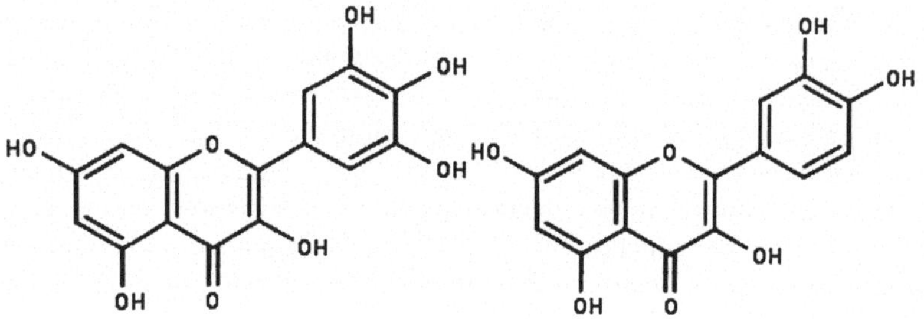

Structures of fisetin and isoliquiritigenin:

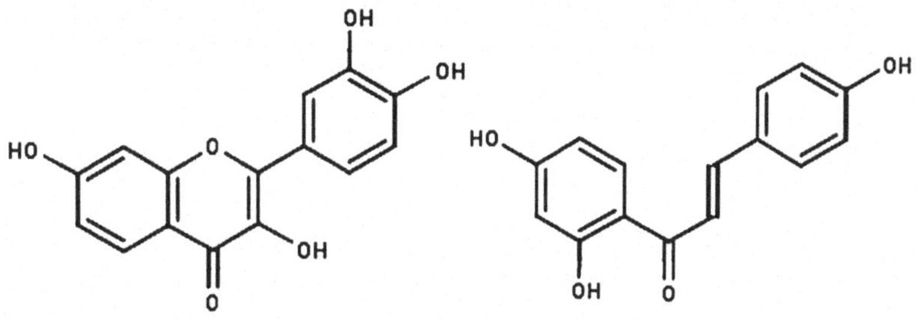

Structures of genistein and daidzein:

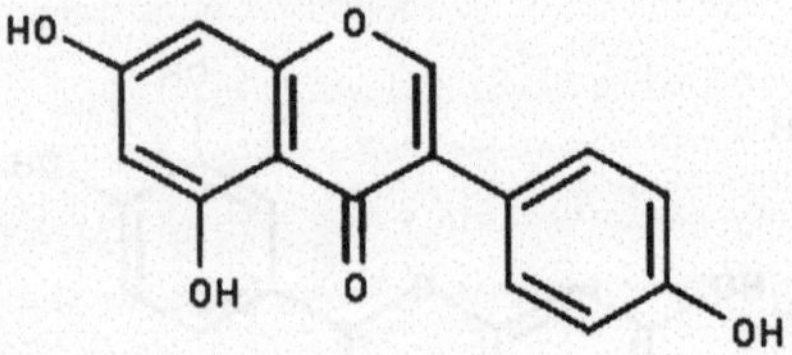

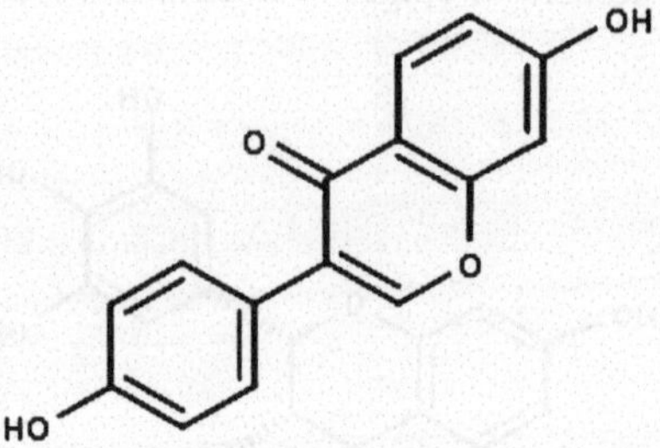

18.4.3 *Isoflavones*

Isoflavones are part of the polyphenol class of phytochemicals related to flavonoids. Genistein, daidzein, and soy isoflavones found in a number of plants and herbs were reported to regulate the expression of genes critical to drug metabolism. Soy isoflavones are a group of compounds isolated from the soybean.

When rats consume a diet high in isoflavone, the activities of kidney glutathione S-transferase and colon quinine reductase are higher. Genistein was found to induce the activity of quinine reductase, while inhibit that of aromatase. The inhibition of aromatase leads to a decrease in estrogen biosynthesis, thus producing an antiestrogenic effect.

Table 18.6 presents the modulation of typical isoflavones on xenobiotic metabolic enzymes. The table reveals that isoflavones induce detoxification enzyme activity, while either inhibit or no effect on CYP450s. These studies were carried out in the experiments using cultured cells from human or rat and animal tissues from rat.

18.5 Diallyl Sulfides

Diallyl sulfide, diallyl disulfide, and diallyl trisulfide are naturally occurring organosulfur compounds. These organosulfur compounds have been recognized as potential chemopreventive agents. The potential mechanisms that are responsible for

Table 18.6 Modulation of isoflavones on xenobiotic metabolic enzymes

	Detoxifying enzyme	Activation enzyme
Modulation compound	Induction	Induction
Genistein	QR	CYP1A1
Daidzein	QR	CYP1A1
Soy isoflavones	GST, QR, and UGT	

Table 18.7 Modulation of diallyl sulfides on xenobiotic metabolic enzymes

	Detoxifying enzymes	Activation enzymes	
Modulation compound	Induction	Induction	Inhibition
Diallyl sulfide	GTS, QOR	CYP1A1, CYP1A2	CYP2E1
Diallyl disulfide	GTS, QOR, UGT		CYP2E1
Diallyl trisulfide	GTS, QOR		

their protective effects against chemically induced cytotoxicity and carcinogenesis are believed to involve metabolic enzyme modulation. Diallyl sulfide, diallyl disulfide, and diallyl trisulfide were all reported to enhance the activity of detoxification enzymes in animal model systems.

Among these organosulfur compounds, the most studied is diallyl sulfide. Diallyl sulfide was found to enhance the activities of QOR and GST in the tissues of the stomach, colon, liver, lung, and bladder. Table 18.7 reports the modulation of diallyl sulfides on xenobiotic metabolic enzymes. These studies were performed in the experiments using animal tissues from rat or mouse or cultured cells from human or animal.

The table reveals that these organosulfur compounds induce detoxification enzymes (GTS, QOR, or UGT), but either inhibit, induce, or no effect on the expression of activation enzymes (CYP450s). The effects of diallyl sulfide on CYP450s are more complicated, since it inhibits CYP2E1 activity but induces CYP1A1 or CYP1A2 activity.

Structures of diallyl sulfide and diallyl disulfide are shown below:

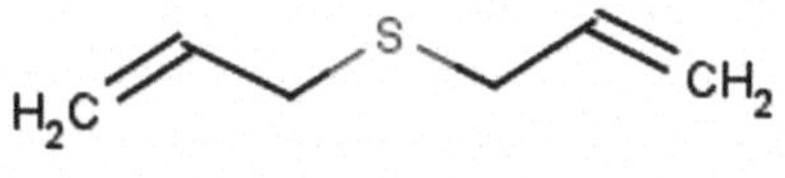

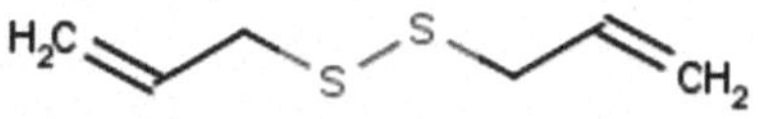

Structure of diallyl trisulfide is also presented here:

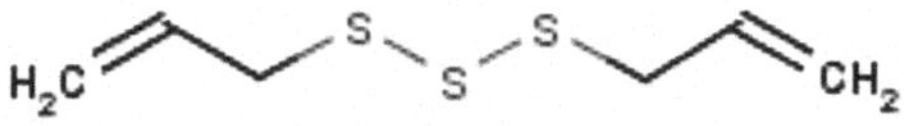

18.6 Terpenes, Terpenoids, and Carotenoids

Terpenes, naturally occurred hydrocarbons, consist of several isoprene units. Among known terpenes are carotene and limonene. Carotene is known for protecting plant cells against destruction from ultraviolet light. Carotene has two primary isomers, α-carotene and ß-carotene, which differ in the position of double bonds in the cyclic group at the end. ß-Carotene, the more common form of carotene, is an antioxidant.

Table 18.8 Modulation of terpenes and terpenoids on xenobiotic metabolic enzymes

	Detoxification enzyme	Activation enzyme
Modulation compound	Induction	Inhibition
ß-Carotene	QOR	
Lycopene	QOR	
Zerumbone	GST	
Canthaxanthin	UGT, QOR	CYP2C19
Astaxanthin	UGT, QOR	CYP2C19

Limonene is abundant in lemon and citrus fruits and is commonly used in perfume, food, and medicines. Limonene is among plant essential oils that have antioxidant and anticancer properties. Terpenoids are a large and diverse class of chemical compounds derived from terpenes. Terpenes and terpenoids are primary components of essential oils of many plants and flowers.

Carotenoids belong to a larger class of terpenoids. The most known carotenoid is lycopene, which is present in ripe fruits, especially tomatoes. Other carotenoids include canthaxanthin, astaxanthin, and zerumbone. Canthaxanthin and astaxanthin are ß-carotene-related compounds. Zerumbone is a phytochemical found in subtropical edible ginger.

Zerumbone exhibits antiproliferative and anti-inflammatory activities and has been implicated as a chemopreventive agent. The exposure of cultured rat liver epithelial cells to zerumbone was found to result in a significant induction of glutathione S-transferase. Zerumbone was also reported to induce nuclear localization of the transcript factor Nrf2 that binds to the antioxidant response element (ARE) of detoxification enzyme genes. This finding suggests that zerumbone is a potential activator of the Nrf2–ARE-dependent detoxification pathway.

Table 18.8 presents the modulation of terpenes, terpenoids, and carotenoids on xenobiotic metabolic enzymes. The table reveals that these compounds exhibit an induction of detoxification enzyme (QOR, GST, or UGT). Meanwhile, canthaxanthin and astaxanthin also exhibit the induction of activation enzyme CYP450. These studies were carried out in the experiments using animal gene and tissue from rat or mouse or cultured cells from human or animal.

Structures of beta-carotene and lycopene and other terpenes or terpenoids are presented below:

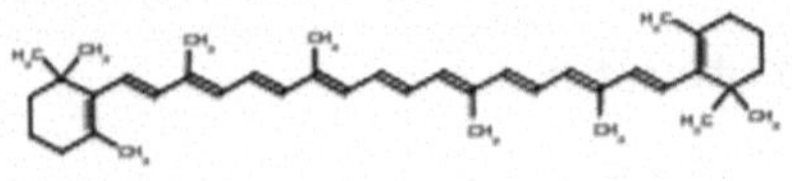

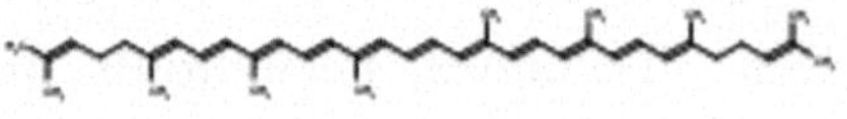

Structures of canthaxanthin and astaxanthin:

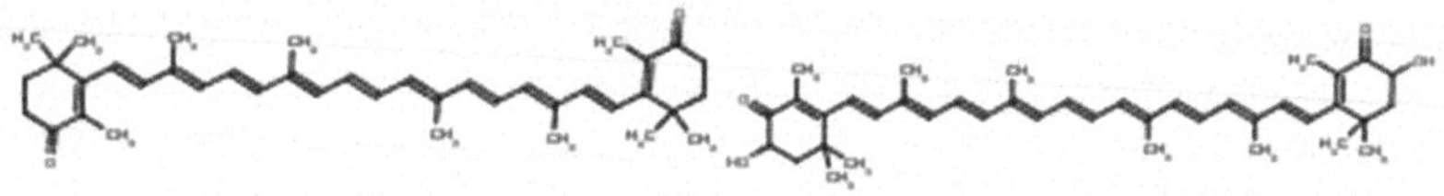

Structures of zerumbone and limonene:

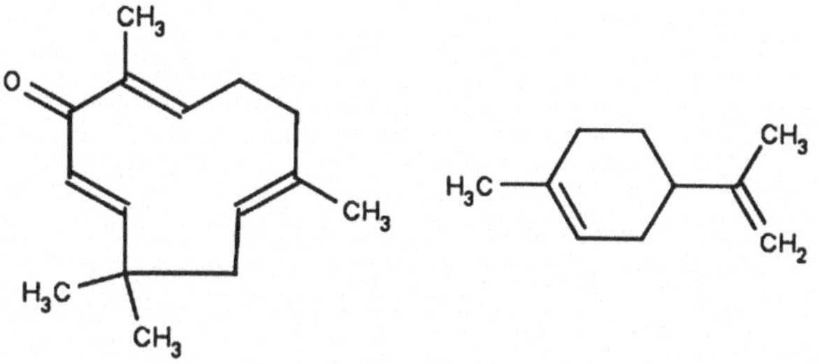

18.7 Other Compounds

Other xenobiotic metabolic enzyme modulators as shown in Table 18.3 include alliin and geniposide. Alliin is the most abundant organosulfur compound in garlic bulbs. Alliin was reported to exhibit the capacity of affecting detoxification enzymes by inducing the expression of GTS or UGT. But, its effect on the expression of CYP450s is more complicated, which involves an inhibition of CYP2E1 or a slight induction of CYP1A2.

Geniposide is an iridoid glycoside isolated from the fruit of *Gardenia jasminoides* Ellis. Geniposide has been shown to exhibit biological capabilities of detoxification, antioxidation, and anticarcinogenesis. Studies of cultured rat hepatocytes have revealed that geniposide possesses a potential for detoxification by inducing GTS and an antioxidative capacity by attenuating oxidative stress-related factors. Geniposide also enhances antioxidant enzyme superoxide dismutase. Moreover, penta-acetyl geniposide was found to decrease aflatoxin B1-induced DNA damage and hepatocarcinogenesis by activating GTS and GSH peroxidase.

18.8 Characteristics of Metabolic Enzyme Modulators

Examinations of the above-discussed inducers of detoxification enzymes reveal that these modulators consist of Michael acceptor functionalities or have the presence of hydroxyl groups at ortho-position on the aromatic rings.

Michael acceptor functionalities are such as olefins and acetylenes. Olefins are unsaturated hydrocarbons that contain carbon-carbon double bond. Acetylenes are unsaturated hydrocarbons that consist of carbon-carbon triple bond. Unsaturated carbon-carbon bonds serve as a source of electrons, whose availability is affected by the groups attached to it.

Moreover, the presence of hydroxyl groups at ortho-position on the aromatic rings is an important structural element for a high inducer potency. Such ortho-hydroxyl groups were found to have a significant impact on the protective role of inducers of detoxification enzymes.

Bibliography

Ahn D, Putt D, Kresty L et al (1996) The effects of dietary ellagic acid on rat hepatic and esophageal mucosal cytochromes P450 and phase II enzymes. Carcinogenesis 17:821–828

Albena T, Dinkova-Kostova MA, Massiah RE, Bozak RJ, Talalay P (2001) Potency of Michael reaction acceptors as inducers of enzymes that protect against carcinogenesis depends on their reactivity with sulfhydryl groups. Proc Natl Acad Sci 98(6):3404–3409

Barry M, Feely J (1990) Enzyme induction and inhibition. Pharmacol Ther 48:71–94

Ben-Dor A, Steiner M, Gheber L et al (2005) Carotenoids activate the antioxidant response element transcription system. Mol Cancer Ther 4:177–186

Brady JF, Ishizaki H, Fukuto JM et al (1991) Inhibition of cytochrome P-450 2E1 by diallyl sulfide and its metabolites. Chem Res Toxicol 4:642–647

Brooks JD, Paton VG, Vidanes G (2001) Potent induction of phase 2 enzymes in human prostate cells by sulforaphane. Cancer Epidemiol Biomarkers Prev 10:949–954

Caro AA, Davis A, Fobare S, Horan N, Ryan C, Schwab C (2019) Antioxidant and pro-oxidant mechanisms of (+) catechin in microsomal CYP2E1-dependent oxidative stress. Toxicol In Vitro 54:1–9

Chen CH (2012) Activation and detoxification enzymes: functions and implications. Springer Sciences, New York

Clark J, You M (2006) Chemoprevention of lung cancer by tea. Mol Nutr Food Res 50:144–151

Crowell PL, Gould MN (1994) Chemoprevention and therapy of cancer by d-limonene. Crit Rev Oncog 5:1–22

Davenport DM, Wargovich MJ (2005) Modulation of cytochrome P450 enzymes by organosulfur compounds from garlic. Food Chem Toxicol 43:1753–1762

Diaz-Gerevini GT, Repossi G, Dain A, Tarres MC, Das UN, Eynard AR (2016) Beneficial action of resveratrol: how and why? Nutrition 32(2):174–178

Dickmann LJ, Vanden Brink BM, Lin YS (2012) In vitro hepatotoxicity and cytochrome P450 induction and inhibition characteristics of carnosic acid, a dietary supplement with antiadipogenic properties. Drug Metab Dispos 40(7):1263–1267

Fahey JW, Talalay P (1999) Antioxidant functions of sulforaphane: a potent inducer of Phase II detoxication enzymes. Food Chem Toxicol 37:973–979

Fernandez-Abascal J, Ripullone M, Valeri A, Leone C, Valoti M (2018) β-naphtoflavone and ethanol induce cytochrome P450 and protect towards MPP+ toxicity in human neuroblastoma SH-SY5Y cells. Int J Mol Sci 19(11):E3369

Fukao T, Hosono T, Misawa S et al (2004) The effects of allyl sulfides on the induction of phase II detoxification enzymes and liver injury by carbon tetrachloride. Food Chem Toxicol 42:743–749

Gradelet S, Astorg P, Leclerc J et al (1996) Effects of canthaxanthin, astaxanthin, lycopene and lutein on liver xenobiotic-metabolizing enzymes in the rat. Xenobiotica 26:49–63

Guyonnet D, Belloir C, Suschetet M et al (2001) Antimutagenic activity of organosulfur compounds from Allium is associated with phase II enzyme induction. Mutat Res 495:135–145

Hamilton SM, Teel RW (1996) Effects of isothiocyanates on cytochrome P-450 1A1 and 1A2 activity and on the mutagenicity of heterocyclic amines. Anticancer Res 16:3597–3602

He T, Shen H, Zhu J, Zhu Y, He Y, Li Z, Lu H (2019) Geniposide attenuates cadmium-induced oxidative stress injury via Nrf2 signaling in osteoblasts. Mol Med Rep 20(2):1499–1508

Hecht SS (2000) Inhibition of carcinogenesis by isothiocyanates. Drug Metab Rev 32:395–411

Kang JJ, Wang HW, Liu TY et al (1997) Modulation of cytochrome P-450-dependent monooxygenases, glutathione and glutathione S-transferase in rat liver by geniposide from Gardenia jasminoides. Food Chem Toxicol 35:957–965

Krajka-Kuźniak V, Szaefer H, Baer-Dubowska W (2004) Modulation of 3-methylcholanthrene-induced rat hepatic and renal cytochrome P450 and phase II enzymes by plant phenols: protocatechuic and tannic acids. Toxicol Lett 152:117–126

Kwak MK, Egner PA, Dolan PM et al (2001) Role of phase 2 enzyme induction in chemoprotection by dithiolethiones. Mutat Res 480–481:305–315

Lançon A, Hanet N, Jannin B et al (2007) Resveratrol in human hepatoma HepG2 cells: metabolism and inducibility of detoxifying enzymes. Drug Metab Dispos 35:699–703

Li Y, Cao Z, Zhu H (2006) Upregulation of endogenous antioxidants and phase 2 enzymes by the red wine polyphenol, resveratrol in cultured aortic smooth muscle cells leads to cytoprotection against oxidative and electrophilic stress. Pharmacol Res 53:6–15

Li Y, Mezei O, Shay NF (2007) Human and murine hepatic sterol-12-alpha-hydroxylase and other xenobiotic metabolism mRNA are upregulated by soy isoflavones. J Nutr 137:1705–1712

Mahéo K, Morel F, Langouët S et al (1997) Inhibition of cytochromes P-450 and induction of glutathione S-transferases by sulforaphane in primary human and rat hepatocytes. Cancer Res 57:3649–3652

Moon YJ, Wang X, Morris ME (2006) Dietary flavonoids: effects on xenobiotic and carcinogen metabolism. Toxicol In Vitro 20:187–210

Munday R, Munday CM (2001) Relative activities of organosulfur compounds derived from onions and garlic in increasing tissue activities of quinone reductase and glutathione transferase in rat tissues. Nutr Cancer 40:205–210

Munday R, Munday CM (2004) Induction of phase II enzymes by 3H-1,2-dithiole-3-thione: dose–response study in rats. Carcinogenesis v25:1721–1725

Munday R, Zhang Y, Paonessa JD et al (2010) Synthesis, biological evaluation, and structure-activity relationships of dithiolethiones as inducers of cytoprotective phase 2 enzymes. J Med Chem 53:4761–4767

Nakajima M, Yoshida R, Shimada N et al (2001) Inhibition and inactivation of human cytochrome P450 isoforms by phenethyl isothiocyanate. Drug Metab Dispos 29:1110–1113

Nakamura Y, Yoshida C, Murakami A et al (2004) a tropical ginger sesquiterpene, activates phase II drug metabolizing enzymes. FEBS Lett 572:245–250

Ow YY, Stupans I (2003) Gallic acid and gallic acid derivatives: effects on drug metabolizing enzymes. Curr Drug Metab 4:241–248

Reicks MM, Crankshaw DL (1996) Modulation of rat hepatic cytochrome P-450 activity by garlic organosulfur compounds. Nutr Cancer 25:241–248

Rodriguez M, Potter DA (2013) CYP1A1 regulates breast cancer proliferation and survival. Mol Cancer Res 11(7):780–792

Roebuck BD, Curphey TJ, Li Y et al (2003) Evaluation of the cancer chemopreventive potency of dithiolethione analogs of oltipraz. Carcinogenesis *24:1919*–1928

Rogan EG (2006) The natural chemopreventive compound indole-3-carbinol: state of the science. In Vivo 20:221–228

Romilly E, Hodges RE, Minich DM (2015) Modulation of metabolic detoxification pathways using foods and food-derived components: a scientific review with clinical application. J Nutr Metabol 2015:760689

S L, Furge LL, Kerriguy N, Nakamura K, Guillouzo A, Guengerich FP (2000) Inhibition of human cytochrome P450 enzymes by 1,2-dithiole-3-thione, oltipraz and its derivatives, and sulforaphane. Chem Res Toxicol 13(4):245–252

Seki H, Akiyoshi T, Imaoka A, Ohtani H (2019) Inhibitory kinetics of fruit components on CYP2C19 activity. Drug Metab Pharmacokinet 34(3):181–186

Seo K, Jung S, Park M et al (2001) Effects of leucocyanidines on activities of metabolizing enzymes and antioxidant enzymes. Biol Pharm Bull 24:592–593

Shamsi S, Tran H, Tan RS, Tan ZJ, Lim LY (2017) Piperine, and Capsaicin: a comparative study of spice-mediated inhibition of human cytochrome P450 isozyme activities. Drug Metab Dispos 45 (1):49–55

Shertzer HG, Puga A, Chang C, Smith P, Nebert DW, Setchell KD, Dalton TP (1999) Inhibition of CYP1A1 enzyme activity in mouse hepatoma cell culture by soybean isoflavones. Chem Biol Interact 123(1):31–49

Shih PH, Yeh CT, Yen GC (2007) Anthocyanins induce the activation of phase II enzymes through the antioxidant response element pathway against oxidative stress-induced apoptosis. J Agric Food Chem 55:9427–9435

Song LL, Kosmeder JW II, Lee SK et al (1999a) Cancer chemopreventive activity mediated by 4'-bromoflavone, a potent inducer of phase II detoxification enzymes. Cancer Res 59:578–585

Song LL, Kosmeder JW II, Lee SK, Gerhäuser C, Lantvit D, Moon RC, Moriarty RM, Pezzuto JM (1999b) Cancer chemopreventive activity mediated by 4'-bromoflavone, a potent inducer of phase II detoxification enzymes. Cancer Res 59(3):578–585

Steele VE, Kelloff GJ, Balentine D et al (2000) Comparative chemopreventive mechanisms of green tea, black tea and selected polyphenol extracts measured by in vitro bioassays. Carcinogenesis 21:63–67

Talalay P (2000) Chemoprotection against cancer by induction of phase 2 enzymes. Biofactors 12 (1-4):5–11

Talalay P, Fahey JW, Holtzclaw WD, Prestera T, Zhang Y (1995) Chemoprotection against cancer by phase 2 enzyme induction. Toxicol Lett 82–83:173–179

Tsai CW, Chen HW, Yang JJ et al (2007) Diallyl disulfide and diallyl trisulfide up-regulate the expression of the pi class of glutathione S-transferase via an AP-1-dependent pathway. J Agric Food Chem 55:1019–1026

Vijaya Padma V, Kalai Selvi P, Sravani S (2014) Protective effect of ellagic acid against TCDD-induced renal oxidative stress: modulation of CYP1A1 activity and antioxidant defense mechanisms. Mol Biol Rep 41(7):4223–4232

von Weymarn LB, Chun JA, Hollenberg PF (2006) Effects of benzyl and phenethyl isothiocyanate on P450s 2A6 and 2A13: potential for chemoprevention in smokers. Carcinogenesis 27 (4):782–790

von Weymarn LB, Chun JA, Knudsen GA et al (2007) Effects of eleven isothiocyanates on P450 2A6- and 2A13-catalyzed coumarin 7-hydroxylation. Chem Res Toxicol 20:1252–1259

Wallig MA, Kingston S, Staack R et al (1998) Induction of rat pancreatic glutathione S-transferase and quinone reductase activities by a mixture of glucosinolate breakdown derivatives found in Brussels sprouts. Food Chem Toxicol 36:365–373

Wang W, Liu LQ, Higuchi CM et al (1998) Induction of NADPH:quinone reductase by dietary phytoestrogens in colonic Colo205 cells. Biochem Pharmacol 56:189–195

Xu M, Dashwood RH (1999) Chemoprevention studies of heterocyclic amine-induced colon carcinogenesis. Cancer Lett 143:179–183

Yannai S, Day AJ, Williamson G et al (1998) Characterization of flavonoids as monofunctional or bifunctional inducers of quinone reductase in murine hepatoma cell lines. Toxicology 36:623–630

Yao HT, Chang YW, Lan SJ, Yeh TK (2008) The inhibitory effect of tannic acid on cytochrome P450 enzymes and NADPH-CYP reductase in rat and human liver microsomes. Food Chem Toxicol 46(2):645–653

Zamaratskaia G, Thøgersen R, Candek-Potokar M, Rasmussen MK (2018) Co-treatment with indole-3- carbinol and resveratrol modify porcine CYP1A and CYP3A activities and expression. Xenobiotica 48(3):232–240

Zhang W, Go ML (2007) Quinone reductase induction activity of methoxylated analogues of resveratrol. Eur J Med Chem 42(6):841–850

Zheng YF, Bae SH, Kwon MJ, Park JB, Choi HD, Shin WG, Bae SK (2013) Inhibitory effects of astaxanthin, β-cryptoxanthin, canthaxanthin, lutein, and zeaxanthin on cytochrome P450 enzyme activities. Food Chem Toxicol 59:78–85

Zhou R, Lin J, Wu D (2014) Sulforaphane induces Nrf2 and protects against CYP2E1-dependent binge alcohol-induced liver steatosis. Biochim Biophys Acta 1840(1):209–218

Chapter 19
Dietary Inducers of Detoxification Enzymes

Carcinogens such as heterocyclic aromatic amines and polycyclic aromatic hydrocarbons can be formed in meat grilled at high temperatures. In metabolism of foreign compounds, activation enzymes catalyze the functionalization of these chemicals to form reactive intermediates or metabolites, followed by detoxification enzyme-catalyzed conjugation reactions, serving to detoxify them and facilitate their excretion from the body.

Oxidative stress and inflammation mediated by reactive chemical intermediates or metabolites, including reactive oxygen species and free radicals, are involved in the aging process and the development of chronic disease conditions. To combat such health effects, one main strategy is to focus on exogenous antioxidants, and another promising approach is to identify phytochemicals derived from vegetables, fruits, and beverages that target cytoprotective mechanisms. The latter approach is primarily concentrated on the induction of the expression of genes that are involved in metabolic detoxification.

Animal, cell culture, and epidemiological studies have demonstrated that consumption of vegetables and fruits is inversely correlated with chronic disease incidence. Besides being the sources of fiber, vitamins, and minerals, vegetables and fruits also contain non-nutritive components such as phytochemicals that can provide substantial health benefits beyond basic nutrition.

Human clinical trials have also been investigated to elucidate phytochemicals present in vegetables and fruits that demonstrate health benefits. Non-nutritive components that offer protective effects for humans against the risk of degenerative and chronic diseases are believed to link to the induction of the expression of metabolic detoxification enzymes.

In addressing to the issue of metabolic detoxification enzyme induction, genetic polymorphisms of these enzymes play an important factor, since they account in part for individual variations in health benefits. Another important factor that requires attention is the inducer-drug interactions. Phytochemicals present in vegetables and fruits may influence pharmacological activity of drugs by modifying their absorption characteristics.

C.-H. Chen, *Xenobiotic Metabolic Enzymes: Bioactivation and Antioxidant Defense*,
https://doi.org/10.1007/978-3-030-41679-9_19

Phytochemicals are thought to induce metabolic detoxification enzymes through nuclear factor E2-related factor 2 (Nrf2)-antioxidant response element (ARE) pathway. Nrf2 binding to ARE in the upstream regulatory region of detoxification genes is essential for the induction of detoxification enzymes. Accordingly, activation of the transcription factor Nrf2 has been proposed as a promising strategy in attenuating oxidative stress and inflammation.

The present chapter discusses dietary inducers of detoxification enzymes. Phytochemicals that are present in vegetables, fruits, green teas, and algae are described below. The potential role of genetic polymorphisms with respect to enzyme modulation was already discussed in the previous chapter.

19.1 Cruciferous Vegetables

Cruciferous vegetables are vegetables of the family Brassicaceae, including broccoli, cauliflower, Brussels sprouts, cabbage and radish. The consumption of these vegetables has long been reported to associate with a reduced risk in the occurrence of degenerative and chronic diseases, such as cardiovascular disease and certain cancers.

The effectiveness of cruciferous vegetables impacts on human health, and cell protection against cytotoxicity and carcinogenesis is dependent on the activities of detoxification enzymes encoded by polymorphic genes. Hence, researchers also take into consideration the impact of metabolic enzyme polymorphisms in exploring the health effects of cruciferous vegetables.

Phytochemical compounds present in cruciferous vegetables that provide protections to living cells against harmful toxicants include sulfur-containing compounds (such as isothiocyanates, sulforaphane, phenethyl isothiocyanate, and 1,2-dithiole-3-thiones) and non-sulfur-containing compounds (such as indole-3-carbinol).

Phytochemicals present in cruciferous, allium, and root vegetables are most investigated for their cytoprotective effects against cytotoxicity and carcinogenesis. Metabolic enzyme modulators that are present in these vegetables are described below:

19.1.1 Sulfur-Containing Compounds

19.1.1.1 Isothiocyanates

Isothiocyanates (ITCs) include sulforaphane and phenethyl isothiocyanate that are broadly distributed among cruciferous vegetables such as cabbages, broccoli, and Brussels sprouts. Isothiocyanates have a chemical structure of R-N=C=S (where R is the aliphatic or aromatic group). Natural ITCs are initially present in cruciferous vegetables as glucosinolates. Enzyme myrosinase converts glucosinolates to isothiocyanates.

Cruciferous vegetables are commonly processed before consumption, where significant alteration in their phytochemical compositions may occur. Heat processing of Brussels sprouts largely reduces the concentrations of isothiocyanates, affecting their ability to induce detoxification enzymes. Moreover, genetic variations and polymorphisms of detoxification enzymes can also affect the effectiveness of isothiocyanates on detoxifying reactive chemical intermediates or metabolites.

The intake of isothiocyanates has been reported to associate with chemoprevention through the induction of detoxification enzymes. Extensive research supports the evidences for the modulation of the mutagenicity of aromatic amines by isothiocyanates.

Moreover, epidemiologic data also support the association of high intake of vegetables with low risk of degeneration and chronic diseases. Experimental dietary studies in humans have further shown the capacity of cruciferous vegetables' constituents to modulate potential disease-preventive mechanisms. Investigations of cytoprotective effects of cruciferous vegetables were primarily carried out in animals and cell or tissue cultures.

19.1.1.2 Sulforaphane

Sulforaphane, the most extensively studied isothiocyanate, is an aliphatic isothiocyanate with a chemical formula of -SO-$(CH_2)_4$-N=C=S. Sulforaphane is commonly found in cruciferous vegetables, especially broccoli. Broccoli accumulates a significant amount of glucoraphanin, the predominant aliphatic glucosinolate, which is metabolized to biologically active sulforaphane.

Exploration into broccoli's impact on human health and cell protection against cytotoxicity and carcinogenesis has been actively pursued. These studies support the association of sulforaphane with the induction of metabolic detoxification enzymes such as glutathione S-transferases and quinone reductase.

19.1.1.3 Phenethyl Isothiocyanate

Another studied isothiocyanate is phenethyl isothiocyanate (PEITC). PEITC is an aromatic isothiocyanate, which is primarily metabolized by glutathione conjugation and is excreted in the urine and bile as the mercapturate. PEITC affects the metabolism of foreign compounds through modulation of both activation enzymes and detoxification enzymes.

PEITC was reported to modulate the metabolism of tobacco-specific carcinogenic nitrosamines by inhibiting its CYP450-mediated bioactivation. PEITC was also found to be a potent inducer of detoxification enzymes such as glutathione S-transferase, UDP-glucuronosyl transferase, and quinone reductase, leading to detoxifying reactive intermediates and metabolites, thus facilitating their excretion from the body.

19.1.1.4 1,2-Dithiole-3-Thione

1,2-Dithiole-3-thione is a cyclic sulfur-containing compound present in cruciferous vegetables. Existing evidences have demonstrated that dithiolethiones are able to efficiently induce the expressions of metabolic detoxification enzymes and antioxidants. The derivatives of 1,2-dithiole-3-thione were also reported to decrease the incidence and multiplicity of tumors in animals exposed to chemical carcinogens.

These findings support the ability of 1,2-dithiole-3-thione and its derivatives to enhance the expression of detoxification enzymes. As a potent enzyme inducer and an anticarcinogen agent, 1,2-dithiole-3-thione was also reported to potently upregulate antioxidant genes in cells by stimulating Nrf2 activity.

In summary, typical sulfur-containing dietary inducers of detoxification enzymes are presented in Table 19.1.

19.1.2 Non-sulfur-Containing Compounds

Unlike the above-discussed isothiocyanates and 1,2-dithiole-3-thione, indole-3-carbinol is a non-sulfur-containing compound and is a hydrolysis product of glucosinolate. Research on indole-3-carbinol has mainly been focused on its cytoprotective effects. Indole-3-carbinol and its dimeric product 3,3-′-diindolylmethane were found to upregulate the expression of detoxification

Table 19.1 Typical dietary inducers of detoxification enzymes

Dietary sources	Induction compounds
Vegetables	
Cruciferous vegetable	Sulfur-containing compounds:
	Isothiocyanates, phenethyl isothiocyanate, sulforaphane, and 1,2-dithiole-3-thione
	Non-sulfur-containing compound:
	Indole-3-carbinol
Allium vegetables	Organosulfur compounds:
	Thiols and diallyl sulfides
Root vegetables	Flavonoids, carotenoids, and curcumin
Fruits	Polyphenols: quercetin, kaempferol, pomegranate, anthocyanin, and procyanidin
	Triterpenes: limonoids, olive oil
Green tea	Polyphenols: epigallocatechin-3-gallate
Algae	Chlorophyll, unsaturated fatty acids, polysaccharides

enzymes, such as glutathione S-transferase (GST). A high level of GST enzyme in human gastrointestinal tissues was found to be inversely correlated with cancer risk.

Typical non-sulfur-containing dietary inducers of detoxification enzymes in cruciferous vegetables are listed in Table 19.1.

19.2 Allium Vegetables

Allium vegetables are high in organosulfur compounds with distinctive flavor and smell. Onions and garlic are the most commonly known allium vegetables. Shallots, leeks, and chives are also members of the allium family. Organosulfur-containing compounds present in allium vegetables include thiols, disulfides, and trisulfides. They are found particularly in garlic, onion, leek, and scallion. A thiol contains a SH functional group. A disulfide contains a -S-S- linkage. Diallyl disulfide is an inhibitor of CYP2E1 enzyme which catalyzes the metabolism of alcohol and drugs.

19.2.1 Thiols

S-Alkyl cysteine sulfoxides are typically found in the Alliaceae family. Onion and garlic have been traditionally used for treating various pathologic conditions. Allium-derived sulfur compounds were found to play a role in cardiovascular protection. Onion and garlic extracts were shown to induce metabolic detoxification enzymes such as glutathione S-transferases and NAD(P)H:oxidoreductase.

19.2.2 Diallyl Sulfides

Diallyl sulfide, diallyl disulfide, and diallyl trisulfide derived from garlic and onion were found to increase the activities of detoxification enzymes, such as quinone reductase and glutathione transferase in rat tissues. Epidemiological studies also indicate that dietary intake of allium vegetables decreases the risk of cancer in humans, owing to their ability to induce the activities of metabolic detoxification enzymes.

Typical inducers of detoxification enzymes including thiols and diallyl sulfides in allium vegetables are also presented in Table 19.1.

19.3 Root Vegetables

Dietary root vegetables include radish, carrot, and ginger. Garlic, onion, and scallion also belong to root vegetables. Root vegetables provide most of flavonoids, carotenoids, and curcumin present in the human diet. There has been a growing interest in these chemical compounds due to their potential health benefit effects.

Flavonoids are the most common group of phenolic compounds in vegetables. Phenolics are compounds containing at least one aromatic ring and at least one -OH group. Carotenoids are colored compounds (yellow, orange, or red). Depending on the presence or absence of oxygen, carotenoids can be divided into xanthophylls and carotenes. Xanthophylls include lutein and zeaxanthin. Lycopene is a carotenoid. Curcumin is the primary curcuminoid in turmeric.

Turmeric curcumin is believed to exhibit anti-inflammatory and antioxidant properties.

19.3.1 Flavonoids

Flavonoids are the most important group of phenolics in diets, which consist of catechin, anthocyanidin, flavone, and flavanol. Their modulation of metabolic detoxification enzymes, such as UDP-glucuronosyl transferase, glutathione S-transferase, and quinone reductase, is considered as a major mechanism of anticarcinogenic effects of flavonoids. Studies of the health effects of flavonoids have revealed that dietary flavonoid-mediated induction of intestinal UDP-glucuronosyl transferase 1A1 may be important for the detoxification of colon carcinogens.

Dietary phenolic derivatives of flavones and isoflavones were also found to be associated with reduced cancer rates. Soybeans are abundant in isoflavones that are known inducers of metabolic detoxification enzymes. Flavonoid-containing soy was found to enhance the activities of UDP-glucuronosyl transferase and quinone reductase, thus offering protective health effects from potentially harmful reactive intermediate species derived from foreign compound metabolism.

Moreover, research evidences have revealed that flavonoids, including fisetin, galangin, quercetin, kaempferol, and genistein, might offer important chemoprevention from sulfation-induced carcinogenesis. Quercetin was shown to increase the transcription of NADH:quinone oxidoreductase. Research findings also suggest that flavonoids stimulate the transcription of detoxification enzymes potentially through Nrf2-ARE-dependent pathway.

19.3.2 *Carotenoids*

Carotenoids are pigments responsible for the red, yellow, or orange color in many vegetables including carrots, tomatoes, and berries. Through acting as antioxidants, carotenoid compounds such as lycopene, lutein, and zeaxanthin are known to offer cytoprotective health benefits for the body against cellular damages.

Among carotenoids, lycopene is known as one of the most effective quenchers of reactive oxygen species. Lutein and zeaxanthin have been shown to improve visual acuity and to slow the progression of age-related macular degeneration. Research evidences suggest that carotenoids may upregulate the expressions of metabolic detoxification enzymes and antioxidants through the activation of Nrf2-dependent pathway.

19.3.3 *Curcumin*

Turmeric is part of the ginger family, and curcumin is the primary curcuminoid in turmeric. Curcumin was reported to induce the expression of glutathione S-transferase. Induction of detoxification enzymes and antioxidants was considered as the potential mechanism underlying the anti-inflammatory and anticancer actions of curcumin.

By inducing heme oxygenase 1 and other detoxifying enzymes in neurons, curcumin was found to protect neurons against various modes of oxidative challenges. Epidemiological studies suggested taking curcumin as a preventive agent against aging and neurodegenerative disorders such as Alzheimer's. However, overdose of curcumin was found to cause toxic effects.

Typical inducers of detoxification enzymes including thiols and diallyl sulfides in allium vegetables as well as flavonoids, carotenoids, and curcumin in root vegetables are also presented in Table 19.1.

19.4 Fruits

The most investigated phytochemicals in fruits for protective effects against cytotoxicity and carcinogenesis are polyphenols and triterpenes. Polyphenols are such as flavonoids which include pomegranate, anthocyanins, and procyanidins. Triterpenes, a group of chemical compounds consisting of three terpenes, are a large group of unsaturated hydrocarbons found in oil of citrus trees. The main triterpenes present in olive oil are oleanolic acid, maslinic acid, and uvaol.

Cytoprotective properties of fruits such as citrus fruits, grapes, berries, and pomegranate have been actively investigated. Phenolic compounds in these fruits were reported to exhibit protective effects on the gastrointestinal tract by enhancing the detoxification of reactive intermediates and metabolites, thus preventing foreign compound metabolism-mediated damages to cellular components such as proteins, lipids, and DNA.

Table 19.1 also presents typical inducers of metabolic detoxification enzymes in fruits, including polyphenols and triterpenes. These phytochemicals that exhibit protective effects against cytotoxicity and carcinogenesis are briefly discussed below:

19.4.1 Polyphenols

The most common group of fruit polyphenolic compounds are flavonoids. Flavonoids in citrus fruits, berries, and grapes are such as eriodyctiol, quercetin, kaempferol, pomegranate, anthocyanin, and procyanidin.

19.4.1.1 Eriodyctiol, Quercetin, Kaempferol, and Pomegranate

Among citrus fruits, eriodyctiol is most investigated for cytoprotective effects that show anti-inflammatory and antioxidation. Eriodyctiol's upregulation of detoxification enzyme heme oxygenase 1 was reported to be beneficial for preventing cardiovascular diseases. Quercetin has diverse biological properties including cytoprotective and antitumorigenic effects. An increase in NAD(P)H:quinone oxidoreductase transcription in response to quercetin suggests that polyphenols can stimulate transcription of metabolic detoxification systems, potentially through ARE-dependent mechanism. The antioxidative properties of quercetin may contribute to cancer chemoprevention.

Kaempferol, another polyphenol found in fruits, was reported to exhibit pharmacological activities, including antioxidant and anti-inflammatory. Epidemiological studies have found a positive correlation between the dietary consumption containing kaempferol and a reduced risk of developing several disorders such as cardiovascular diseases and cancers.

Pomegranate in juice and peel are also rich in polyphenols. Pomegranate is gaining tremendous attention due to its powerful antioxidant properties. Mechanistic studies revealed that pomegranate elevates gene expression of hepatic detoxification enzymes and antioxidants. Oxidative stress predisposed to hepatocarcinogenesis is the major driving force of hepatocellular carcinoma. Research has provided substantial evidence to propose that pomegranate-mediated chemoprevention of hepatocarcinogenesis is associated with Nrf2-regulated antioxidant mechanisms.

19.4.1.2 Anthocyanin and Procyanidin

Anthocyanins are also a type of flavonoids. Blueberry anthocyanins offer cytoprotective effects due to their abilities to inhibit oxidative stress and cell proliferation. Effects of anthocyanin fractions from cultivars of blueberries on detoxification enzymes, such as glutathione S-transferase and quinone reductase, are the potential mechanism through which anthocyanins exhibit cytoprotective effects. Anthocyanins in grapes were also found to have breast cancer chemoprevention potential, in part due to their capacity to block carcinogen-DNA adduct formation and to modulate the activities of carcinogen-metabolizing enzymes.

Procyanidins, polymeric flavan-3-ols, are members of the tannin class of flavonoids.

Procyanidins are known to possess antioxidant and anticarcinogenic properties. Treatments of procyanidins were found to significantly induce the expressions of antioxidants such as catalase and superoxide dismutase, as well as the activities of metabolic detoxification enzymes such as quinone oxidoreductase and glutathione peroxidase. Procyanidins from wild grape seeds were used as a potential chemoprevention agent that induces metabolic detoxifying enzymes through the Nrf2-ARE pathway.

19.4.2 Triterpenes

Research on the bioactive properties of phytochemicals revealed that triterpenes present in olive oil are a natural source of antioxidants that could be beneficial for preventing diseases related to oxidative stress damages. Citrus limonoids, highly oxidized triterpenes, may provide substantial anticancer activity. It has been reported that consumption of citrus fruits may get health benefit from limonoids, which were found to influence the activity of metabolic detoxification enzymes, such as glutathione S-transferase in the intestine and NAD(P)H:quinone reductase in the stomach.

19.5 Green Tea

Green tea contains high concentrations of polyphenols that are powerful antioxidants with anticarcinogenic properties. Due to five times higher concentration of epigallocatechin-3-gallate, green tea is assumed to possess a greater preventive potential than black tea. Green tea polyphenols, especially epigallocatechin-3-gallate, were found to modulate metabolic detoxification enzymes, such as glutathione S-transferase, glutathione peroxidase, glutamate cysteine ligase, and heme oxygenase 1, that are involved in the detoxification of reactive chemical intermediates including reactive oxygen species. Besides inducing detoxification enzymes, green tea polyphenols were also shown to exert cancer-protective activity by inhibiting the metabolic activation of carcinogens, particularly CYP450.

Tea polyphenols were also reported to protect against tumors of the lung, gastrointestinal tract, and liver. An inverse association was found between green tea intake and risk of esophageal cancer. Molecular mechanisms underlying chemoprevention effects exerted by green tea have been extensively investigated. Activation of Nrf2 is considered as an important molecular target of chemoprotective agents. Nevertheless, it is essential to point out that the balance between the induction of detoxification enzymes and the inhibition of activation enzymes is important in determining the chemoprevention effects of green tea polyphenols. Table 19.1 also presents green tea in the list of dietary inducers of detoxification enzymes.

19.6 Algae

Unsaturated fatty acid, polysaccharides, and other extracts isolated from edible algae have attracted interests due to their potential biological activities. Algae contain a powerful nutrient called chlorophyll, the key photosynthetic compound present in plants. Methanolic extract of seaweed was shown to revert dioxin-induced toxicity through an alteration in the expression of antioxidants (catalase, superoxide dismutase, glutathione, and glutathione peroxidase). Chlorophyll was also reported for its ability to help cleansing blood from toxins and unwanted materials.

Moreover, algal species (red and brown algae) were found to modulate the expression of detoxification enzyme glutathione transferase against the substrate, aryl halides. Unsaturated fatty acid derivatives isolated as active components from green alga were shown to induce the expression of NAD(P)H:quinone oxidoreductase, heme oxygenase 1, and thioredoxin reductase. These findings led to propose that unsaturated fatty acid derivatives from green alga are able to activate the transcription factor that recognizes the antioxidant response element (ARE).

Exploration of the chemoprevention activity of water-soluble polysaccharide extract revealed that polysaccharide extract from brown alga inhibits the carcinogen activator CYP4501A and induces the carcinogen detoxification enzyme glutathione S-transferase. However, studies of the effects of lipopolysaccharides produced by cyanobacteria (blue-green algae) on detoxification enzymes showed a significant reduction in glutathione S-transferase in embryos exposed to cyanobacterial lipopolysaccharides. Further investigations to explore algae's health benefits beyond basic nutrition are needed.

Table 19.1 also presents chlorophyll, unsaturated fatty acids, and lipopolysaccharides isolated from algae as inducers of detoxification enzymes.

19.7 Inducer-Drug Interactions

Bioavailability is referred to as the degree and rate at which an administered drug is absorbed by the body. Enzyme induction or inhibition that increases or decreases the metabolism of drugs may affect the rate and extent of medications. Knowledge of potent inhibition and induction of drugs can also help minimize the potential adverse drug interactions.

CYP450 enzymes are the major activation enzymes essential for the metabolism of many drugs. CYP450s can be inhibited or induced by drugs, resulting in clinically significant interactions that can cause unanticipated adverse reactions. In a study of interaction between grapefruit juice and felodipine (a dihydropyridine calcium channel antagonist), grapefruit juice was found to reduce the metabolism of felodipine through selective posttranslational downregulation of the expression of CYP3A4 in the intestinal wall.

There are concerns that antioxidative properties of sulforaphane may interfere with cytotoxic drugs such as doxorubicin. Doxorubicin is a cardiotoxic anticancerous drug that is widely used for a broad range of cancers. Sulforaphane was found to protect the heart from doxorubicin-induced toxicity and damage and enhance the efficacy of doxorubicin in regression of tumor growth. The dosage of doxorubicin required to treat the tumor could be reduced when sulforaphane is administered simultaneously. Moreover, sulforaphane was found to potentiate the drug effect in prostate cancer stem cells.

Curcumin is another compound that has been investigated for inducer-drug interaction. Curcumin has the potential to be utilized for chemoprevention and chemotherapy due to its antioxidant, anti-inflammatory, and anticarcinogenic properties. Curcumin has been studied for its synergistic effect on cardiomyocytes and cancer cells. Moreover, dose-dependent cardiotoxicity of doxorubicin can be significantly minimized by supplementing it with curcumin.

The regulation of polyamine metabolism in human gastric and colon carcinoma cell lines in response to curcumin was also investigated to provide insight into the polyamine-related mechanisms involved in the cancer cell in response to curcumin and its potential as a chemopreventive or chemotherapeutic agent in the gastrointestinal tract. Understanding such inducer-drug interaction requires further research to explore the potential of utilizing metabolic enzyme inducers for chemoprevention.

Bibliography

Amararathna M, Johnston MR, Rupasinghe HP (2016) Plant polyphenols as chemopreventive agents for lung cancer. Int J Mol Sci 19(17):8

Angeloni C, Leoncini E, Malaguti M, Angelini S, Hrelia P, Hrelia S (2009) Modulation of phase II enzymes by sulforaphane: implications for its cardioprotective potential. J Agric Food Chem 57 (12):5615–5622

Appelt LC, Reicks MM (1997) Soy feeding induces phase II enzymes in rat tissues. Nutr Cancer 28 (3):270–275

Bailey DG, Malcolm J, Arnold O, Spence JD (1998) Grapefruit juice–drug interactions. Br J Clin Pharmacol 46(2):101–110

Bak MJ, Jun M, Jeong WS (2012) Procyanidins from wild grape (Vitis amurensis) seeds regulate ARE- mediated enzyme expression via Nrf2 coupled with p38 and PI3K/Akt pathway in HepG2 cells. Int J Mol Sci 13(1):801–818

Barry M, Feely J (1990) Enzyme induction and inhibition. Pharmacol Ther 48:71–94

Bishayee A, Bhatia D, Thoppil RJ, Darvesh AS, Nevo E, Lansky EP (2011) Pomegranate-mediated chemoprevention of experimental hepatocarcinogenesis involves Nrf2-regulated antioxidant mechanisms. Carcinogenesis 32(6):888–896

Bolling BW, Parkin KL (2008) Phenolic derivatives from soy flour ethanol extract are potent in vitro quinone reductase (QR) inducing agents. J Agric Food Chem 56(22):10473–10480

Bose C, Awasthi S, Sharma R, Benes H, Hauer-Jensen M, Boerma M, Singh SP (2018) Sulforaphane potentiates anticancer effects of doxorubicin and attenuates its cardiotoxicity in a breast cancer model. PLoS One 13(3):e0193918

Chen CH (2012) Activation and detoxification enzymes: functions and implications. Springer Sciences, New York

Chow HH, Hakim IA, Vining DR, Crowell JA, Tome ME, Ranger-Moore J, Cordova CA, Mikhael DM, Briehl MM, Alberts DS (2007) Modulation of human glutathione s-transferases by polyphenone intervention. Cancer Epidemiol Biomarkers Prev 16(8):1662–1666

Galijatovic A, Otake Y, Walle UK, Walle T (2001) Induction of UDP-glucuronosyltransferase UGT1A1 by the flavonoid chrysin in Caco-2 cells--potential role in carcinogen bioinactivation. Pharm Res 18(3):374–379

Gamal-Eldeen AM, Ahmed EF, Abo-Zeid MA (2009) In vitro cancer chemopreventive properties of polysaccharide extract from the brown alga, Sargassum latifolium. Food Chem Toxicol 47 (6):1378–1384

Guo Z, Smith TJ, Wang E, Sadrieh N, Ma Q, Thomas PE, Yang CS (1992) Effects of phenethyl isothiocyanate, a carcinogenesis inhibitor, on xenobiotic-metabolizing enzymes and nitrosamine metabolism in rats. Carcinogenesis 13(12):2205–2210

Herve C, de Franco PO, Groisillier A, Tonon T, Boyen C (2008) New members of the glutathione transferase family discovered in red and brown algae. Biochem J 412(3):535–544

Hodges RE, Minich DM (2015) Modulation of metabolic detoxification pathways using foods and food-derived components: a scientific review with clinical application. J Nutr Metab 2015:760689

Ioannides C, Konsue N (2015) A principal mechanism for the cancer chemopreventive activity of phenethyl isothiocyanate is modulation of carcinogen metabolism. Drug Metab Rev 47 (3):356–373

Jain A, Rani V (2018) Assessment of herb-drug synergy to combat doxorubicin induced cardiotoxicity. Life Sci 205:97–106

Jaja-Chimedza A, Gantar M, Mayer GD, Gibbs PD, Berry JP (2012) Effects of cyanobacterial lipopolysaccharides from microcystis on glutathione-based detoxification pathways in the zebrafish (Danio rerio) embryo. Toxins (Basel) 4(6):390–404

Kallifatidis G, Labsch S, Rausch V, Mattern J, Gladkich J, Moldenhauer G, Büchler MW, Salnikov AV, Herr I (2011) Sulforaphane increases drug-mediated cytotoxicity toward cancer stem-like cells of pancreas and prostate. Mol Ther 19(1):188–195

Kensler TW, Curphey TJ, Maxiutenko Y, Roebuck BD (2000) Chemoprotection by organosulfur inducers of phase 2 enzymes: dithiolethiones and dithiins. Drug Metabol Drug Interact 17:3–22

Keum YS (2011) Regulation of the Keap1/Nrf2 system by chemopreventive sulforaphane: implications of posttranslational modifications. Ann N Y Acad Sci 1229:184–189

Keum YS, Jeong WS, Kong AN (2005) Chemopreventive functions of isothiocyanates. Drug News Perspect 18(7):445–451

Kwak MK, Egner PA, Dolan PM, Ramos-Gomez M, Groopman JD, Itoh K, Yamamoto M, Kensler TW (2001) Role of phase 2 enzyme induction in chemoprotection by dithiolethiones. Mutat Res 480–481:305–315

Kwak MK, Itoh K, Yamamoto M, Kensler TW (2002) Enhanced expression of the transcription factor Nrf2 by cancer chemopreventive agents: role of antioxidant response element-like sequences in the nrf2 promoter. Mol Cell Biol 22(9):2883–2892

Lampe JW (2007) Diet, genetic polymorphisms, detoxification, and health risks. Altern Ther Health Med 13(2):S108–S111

Lampe JW, Chen C, Li S, Prunty J, Grate MT, Meehan DE, Barale KV, Dightman DA, Feng Z, Potter JD (2000) Modulation of human glutathione S-transferases by botanically defined vegetable diets. Cancer Epidemiol Biomarkers Prev 8:787–793

Lee JM, Johnson JA (2004) An important role of Nrf2-ARE pathway in the cellular defense mechanism. J Biochem Mol Biol 37:139–143

Lee SE, Yang H, Son GW, Park HR, Park CS, Jin YH, Park YS (2015) Eriodictyol protects endothelial cells against oxidative stress-induced cell death through modulating ERK/Nrf2/ARE-dependent Heme Oxygenase-1 expression. Int J Mol Sci 16(7):14526–14539

Leoncini E, Malaguti M, Angeloni C, Motori E, Fabbri D, Hrelia S (2011) Cruciferous vegetable phytochemical sulforaphane affects phase II enzyme expression and activity in rat cardiomyocytes through modulation of Akt signaling pathway. J Food Sci 76:H175–H181

Maliakal P, Sankpal UT, Basha R, Maliakal C, Ledford A, Wanwimolruk S (2011) Relevance of drug metabolizing enzyme activity modulation by tea polyphenols in the inhibition of esophageal tumorigenesis. Med Chem 7(5):480–487

Mastaloudis A, Wood SM (2012) Age-related changes in cellular protection, purification, and inflammation-related gene expression: role of dietary phytonutrients. Ann N Y Acad Sci 1259:112–120

Metere A, Giacomelli L (2017) Absorption, metabolism and protective role of fruits and vegetables polyphenols against gastric cancer. Eur Rev Med Pharmacol Sci 21(24):5850–5858

Moon YJ, Wang X, Morris ME (2006) Dietary flavonoids: effects on xenobiotic and carcinogen metabolism. Toxicol In Vitro 20(2):187–210

Morimitsu Y, Nakagawa Y, Hayashi K, Fujii H, Kumagai T, Nakamura Y, Osawa T, Horio F, Itoh K, Iida K, Yamamoto M, Uchida K (2002) A sulforaphane analogue that potently activates the Nrf2- dependent detoxification pathway. J Biol Chem 277:3456–3463

Munday R, Munday CM (2001) Relative activities of organosulfur compounds derived from onions and garlic in increasing tissue activities of quinone reductase and glutathione transferase in rat tissues. Nutr Cancer 40(2):205–210

Munday R, Munday JS, Munday CM (2003) Comparative effects of mono-, di-, tri-, and tetrasulfides derived from plants of the Allium family: redox cycling in vitro and hemolytic activity and Phase 2 enzyme induction in vivo. Free Radic Biol Med 34(9):1200–1211

Munday R, Zhang Y, Munday CM, Li J (2006) Structure-activity relationships in the induction of Phase II enzymes by derivatives of 3H-1,2-dithiole-3-thione in rats. Chem Biol Interact 160 (2):115–122

Murray-Stewart T, Dunworth M, Lui Y, Giardiello FM, Woster PM, Casero RA Jr (2018) Curcumin mediates polyamine metabolism and sensitizes gastrointestinal cancer cells to antitumor polyamine- targeted therapies. PLoS One 13(8):e0202677

Na HK, Surh YJ (2008) Modulation of Nrf2-mediated antioxidant and detoxifying enzyme induction by the green tea polyphenol EGCG. Food Chem Toxicol 46(4):1271–1278

Nishinaka T, Ichijo Y, Ito M, Kimura M, Katsuyama M, Iwata K, Miura T, Terada T, Yabe-Nishimura C (2007) Curcumin activates human glutathione S-transferase P1 expression through antioxidant response. Toxicol Lett 170:238–247

Noratto GD, Chew BP, Atienza LM (2017) Red raspberry (Rubus idaeus L.) intake decreases oxidative stress in obese diabetic (db/db) mice. Food Chem 227:305–314

Osawa T, Kato Y (2005) Protective role of antioxidative food factors in oxidative stress caused by hyperglycemia. Ann N Y Acad Sci 1043:440–451

Perez JL, Jayaprakasha GK, Valdivia V, Munoz D, Dandekar DV, Ahmad H, Patil BS (2009) Limonin methoxylation influences the induction of glutathione S-transferase and quinone reductase. J Agric Food Chem 57(12):5279–5286

Perez JL, Jayaprakasha GK, Cadena A, Martinez E, Ahmad H, Patil BS (2010) In vivo induction of phase II detoxifying enzymes, glutathione transferase and quinone reductase by citrus triterpenoids. BMC Complement Altern Med 10:51

Rose P, Whiteman M, Moore PK, Zhu YZ (2005) Bioactive S-alk(en)yl cysteine sulfoxide metabolites in the genus Allium: the chemistry of potential therapeutic agents. Nat Prod Rep 22(3):351–368

Saw CL, Guo Y, Yang AY, Paredes-Gonzalez X, Ramirez C, Pung D, Kong AN (2014) The berry constituents quercetin, kaempferol, and pterostilbene synergistically attenuate reactive oxygen species: involvement of the Nrf2-ARE signaling pathway. Food Chem Toxicol 72:303–311

Scapagnini G, Caruso C, Calabrese V (2010) Therapeutic potential of dietary polyphenols against brain ageing and neurodegenerative disorders. Adv Exp Med Biol 698:27–35

Shih PH, Yeh CT, Yen GC (2007) Anthocyanins induce the activation of phase II enzymes through the antioxidant response element pathway against oxidative stress-induced apoptosis. J Agric Food Chem 55:9427–9435

Singletary KW, Jung KJ, Giusti M (2007) Anthocyanin-rich grape extract blocks breast cell DNA damage. J Med Food 10(2):244–251

Srivastava A, Akoh CC, Fischer J, Krewer G (2007) Effect of anthocyanin fractions from selected cultivars of Georgia-grown blueberries on apoptosis and phase II enzymes. J Agric Food Chem 55(8):3180–3185

Syed DN, Chamcheu J-C, Adhami VM, Mukhtar H (2013) Pomegranate extracts and cancer prevention: molecular and cellular activities. Anticancer Agents Med Chem 13(8):1149–1161

Tan XL, Spivack SD (2009) Dietary chemoprevention strategies for induction of phase II xenobiotic- metabolizing enzymes in lung carcinogenesis: a review. Lung Cancer 65:129–137

Truong VL, Bak MJ, Jun M, Kong AN, Ho CT, Jeong WS (2014) Antioxidant defense and hepatoprotection by procyanidins from almond (Prunus amygdalus) skins. J Agric Food Chem 62(34):8668–8678

Valerio LG Jr, Kepa JK, Pickwell GV, Quattrochi LC (2001) Induction of human NAD(P)H: quinone oxidoreductase (NQO1) gene expression by the flavonol quercetin. Toxicol Lett 119 (1):49–57

Wang R, Paul VJ, Luesch H (2013) Seaweed extracts and unsaturated fatty acid constituents from the green alga Ulva lactuca as activators of the cytoprotective Nrf2-ARE pathway. Free Radic Biol Med 57:141–153

Wargovich MJ, Cunningham JE (2003) Diet, individual responsiveness and cancer prevention. J Nutr 133(7 Suppl):2400S–2403S

Wu J-C, Lai C-S, Tsai M-L, Ho C-T, Wang Y-J, Pan M-H (2017) Chemopreventive effect of natural dietary compounds on xenobiotic-induced toxicity. J Food Drug Anal 25(1):176–186

Yang J, Liu RH (2009) Induction of phase II enzyme, quinone reductase, in murine hepatoma cells *in vitro* by grape extracts and selected phytochemicals. Food Chem v114:898–904

Yuan JM (2011) Green tea and prevention of esophageal and lung cancers. Mol Nutr Food Res 55 (6):886–904

Zarfeshany A, Asgary S, Javanmard SH (2014) Potent health effects of pomegranate. Adv Biomed Res 3:100

Correction to: Xenobiotic Metabolic Enzymes: Bioactivation and Antioxidant Defense

Correction to:
C.-H. Chen, ***Xenobiotic Metabolic Enzymes: Bioactivation and Antioxidant Defense,***
https://doi.org/10.1007/978-3-030-41679-9

After initial publication of the book, various errors were identified that needed correction. All corrections listed below have been updated within the current version.

Chapter 4

1. Page 37, Section 4.1.1(f)

Depiction in book as

```
-Φ-RC=CR'   →   -Φ-RC-CHR'
                   \/
                   O
```

Correct depiction:

```
-Φ-RC=CR'   →   -Φ-RC-CHR'
                   \ /
                    O
```

The updated online version of these chapters can be found at
https://doi.org/10.1007/978-3-030-41679-9_4
https://doi.org/10.1007/978-3-030-41679-9_6
https://doi.org/10.1007/978-3-030-41679-9_15

C.-H. Chen, *Xenobiotic Metabolic Enzymes: Bioactivation and Antioxidant Defense*,
https://doi.org/10.1007/978-3-030-41679-9_20

Chapter 6

2. Page 61, Section 6.1.1(e)

Depiction in book as

$$\underset{\displaystyle NH_2}{\overset{}{R-C=O}} + O_2 + NADPH + H^+ \rightarrow RCHO + NH_3OH + H_2O + NADP^+$$

Correct depiction:

$$R-\underset{\displaystyle NH_2}{\underset{|}{C}}=O + O_2 + NADPH + H^+ \rightarrow RCHO + NH_3OH + H_2O + NADP^+$$

Depiction in book as

$$\underset{\displaystyle OH}{\overset{}{R-C=O}} + O_2 + NADPH + H^+ \rightarrow HCHO + HCOOH + H_2O + NADP^+$$

Correct depiction:

$$R-\underset{\displaystyle OH}{\underset{|}{C}}=O + O_2 + NADPH + H^+ \rightarrow HCHO + HCOOH + H_2O + NADP^+$$

3. Page 66, Section 6.2.2(a)

Depiction in book as

$$2\ \underset{\displaystyle R'}{\underset{|}{R}}\text{-C=N-H} + O_2 \rightarrow 2\ \underset{\displaystyle R'}{\underset{|}{R}}\text{-C=N-OH}$$

Correct depiction:

$$2\ R\text{-}\underset{\displaystyle R'}{\underset{|}{C}}\text{=N-H} + O_2 \rightarrow 2\ R\text{-}\underset{\displaystyle R'}{\underset{|}{C}}\text{=N-OH}$$

4. Page 67, Section 6.2.3

Depiction in book as

$$R-C=C-R' + O_2 \rightarrow \overset{O}{/\backslash} \quad R-C-C-R'$$

Correct depiction

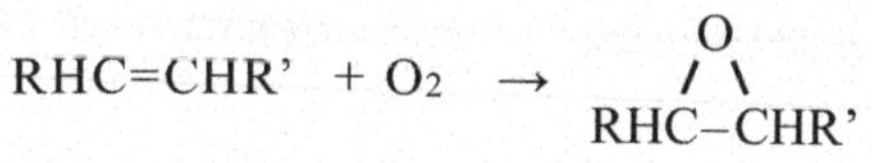

Depiction in book as

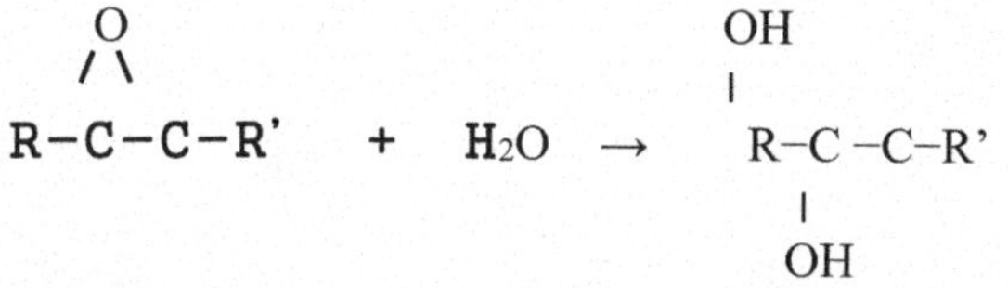

Correct depiction:

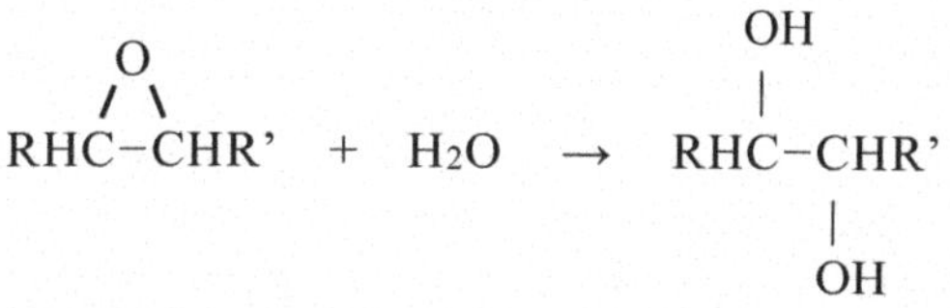

5. Page 68, Section 6.3.2

Depiction in book as

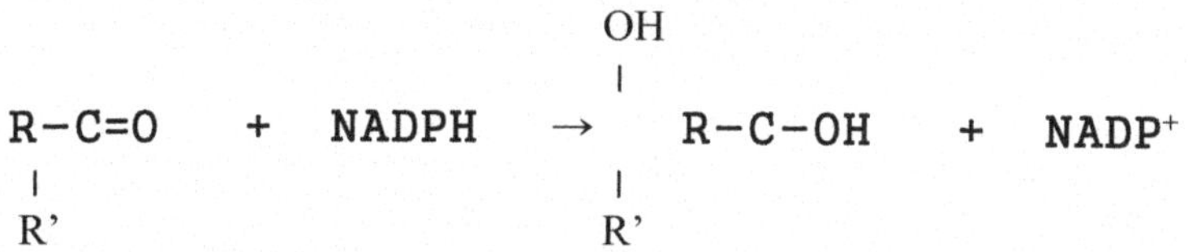

Correct depiction:

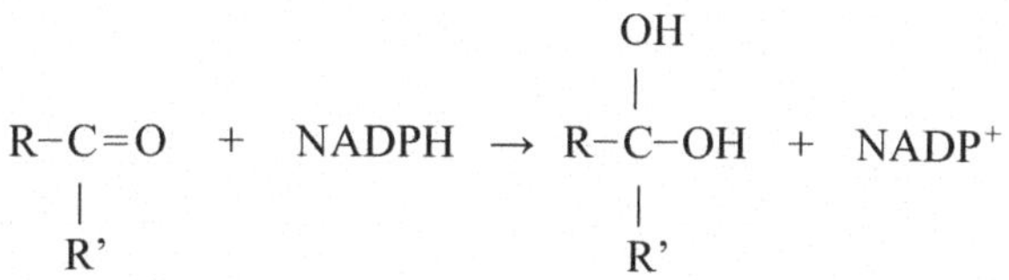

6. Page 68, Section 6.4.1

Depiction in book as

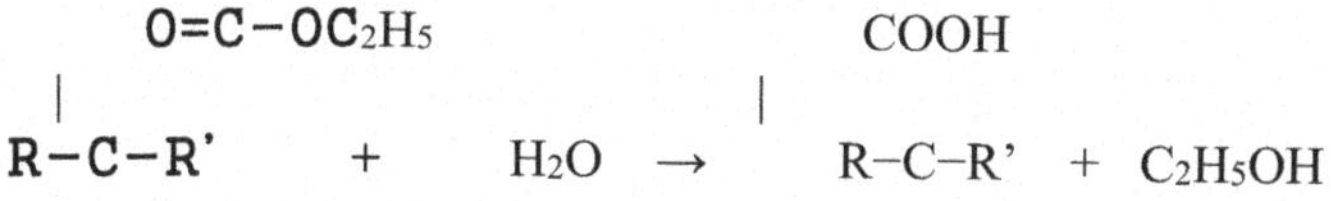

Correct depiction:

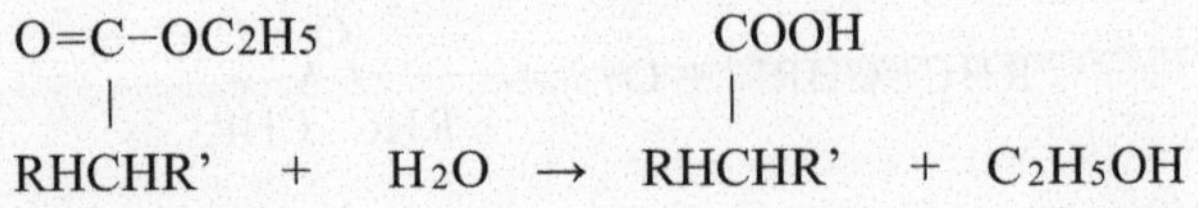

7. Page 69, Section 6.4.2

Depiction in book as

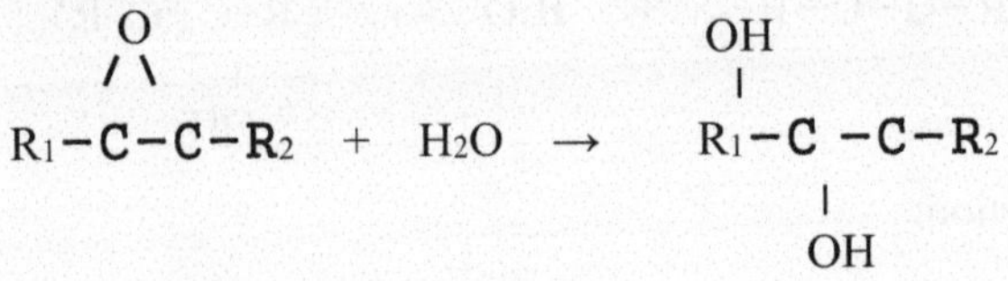

Correct depiction:

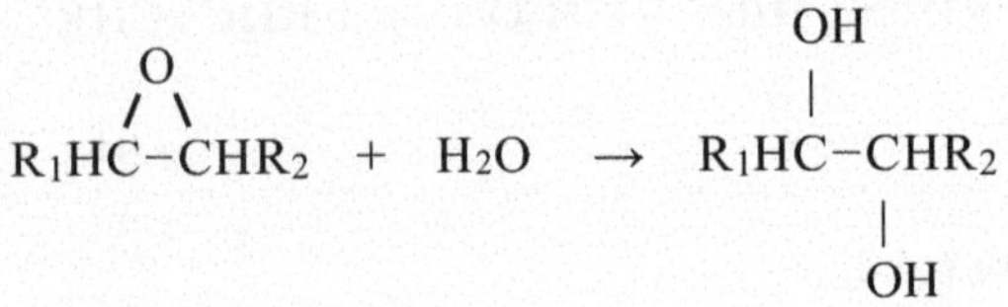

Depiction in book as

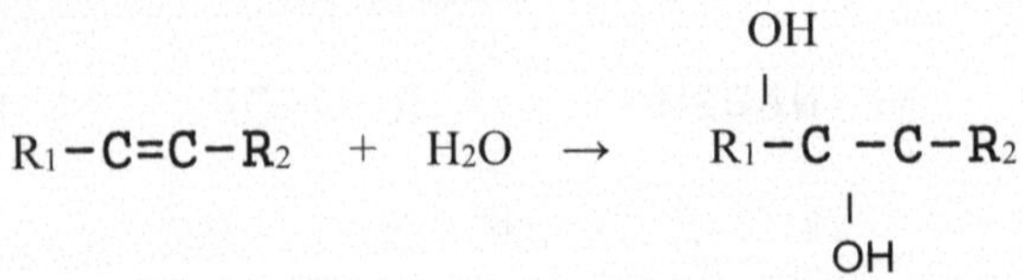

Correct depiction:

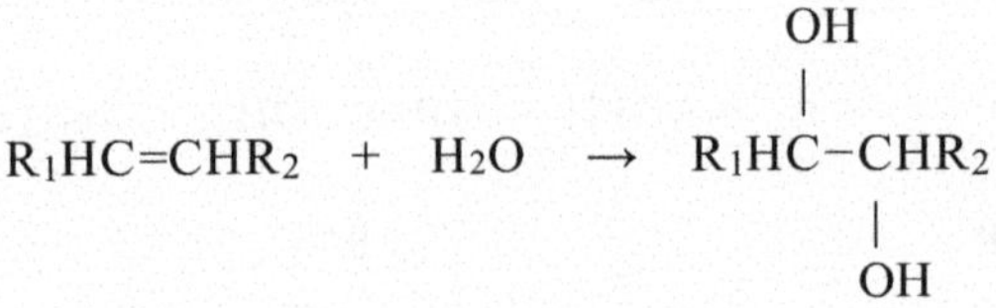

Chapter 15

8. Page 179, Figure 15.7

Depiction in book as

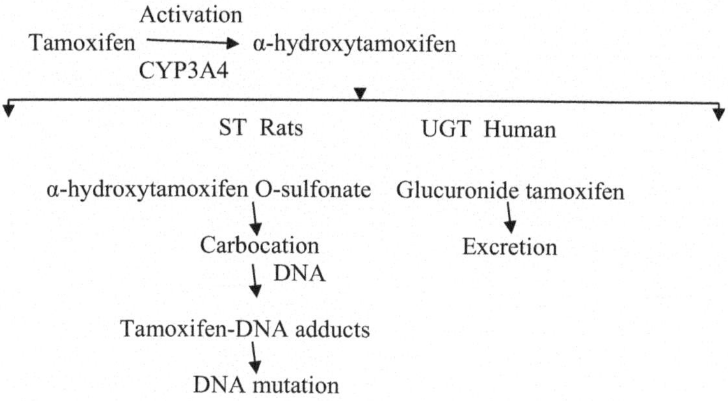

Correct depiction:

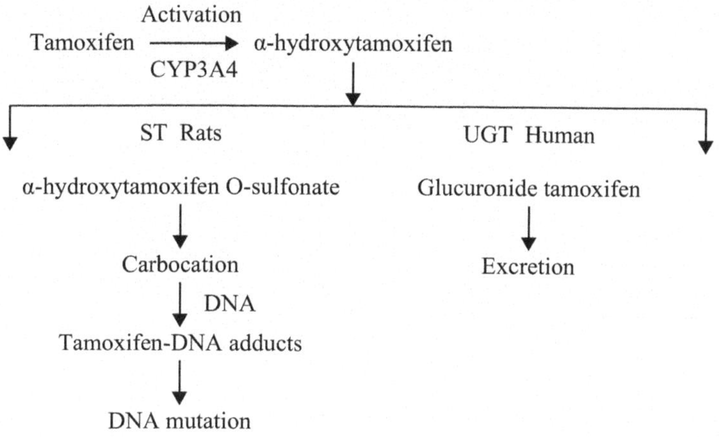

Index

C.-H. Chen, *Xenobiotic Metabolic Enzymes: Bioactivation and Antioxidant Defense*,
https://doi.org/10.1007/978-3-030-41679-9